Sencha Touch in Action

Sencha Touch in Action

JESUS GARCIA
ANTHONY DE MOSS
MITCHELL SIMOENS

MANNING
SHELTER ISLAND

For online information and ordering of this and other Manning books, please visit www.manning.com. The publisher offers discounts on this book when ordered in quantity. For more information, please contact

Special Sales Department
Manning Publications Co.
20 Baldwin Road
PO Box 261
Shelter Island, NY 11964
Email: orders@manning.com

Photographs in this book were created by Martin Evans and Jordan Hochenbaum, unless otherwise noted. Illustrations were created by Martin Evans, Joshua Noble, and Jordan Hochenbaum. Fritzing (fritzing.org) was used to create some of the circuit diagrams.

Manning Publications Co.
20 Baldwin Road
PO Box 261
Shelter Island, NY 11964

Development editor: Frank Pohlmann
Technical proofreader: Jamund Ferguson
Copyeditor: Liz Welch
Proofreader: Melody Dolab
Typesetter: Dennis Dalinnik
Cover designer: Marija Tudor

ISBN: 9781617290374
Printed in the United States of America
1 2 3 4 5 6 7 8 9 10 – MAL – 19 18 17 16 15 14 13

To my wife Erika and my sons Takeshi and Kenji

—J.G.

brief contents

contents

preface

I started my career in the world of Sencha back in 2006 when the precursor to what's known as ExtJS today (Sencha's desktop JavaScript framework) was still an experiment. Soon after my introduction, I became addicted to the design patterns that were promoted by the quickly evolving framework. But more importantly, I fell in love with the thriving community of developers looking to contribute.

I was inspired by many of the active members and decided to become a contributing member myself, spending tens of hours per week answering people's questions, writing blog posts, and eventually publishing instructional screencasts.

I can remember the feeling of excitement when I learned of the development of Sencha Touch back in 2009. Back then, HTML5 was something that people didn't talk about much, and it was only a draft specification. In 2012, Sencha Touch 1.0 arrived and was the first HTML5 mobile framework on the market. A lot of the technology driving Sencha Touch is considered revolutionary in the industry. For example, the Sencha Touch scroller is one of the most robust and smoothest available for single-page applications for mobile devices.

To be honest, I'd never done mobile development before then. I didn't own a smartphone and I didn't know much about WebKit. I'm happy to say that Sencha Touch changed all that for me!

The reason I got into mobile development was because Sencha Touch allowed me to use my existing HTML, JavaScript, and CSS skills to build applications for smartphones. Back in 2009, I didn't have time to learn Objective C and I honestly didn't care much for Java.

Fast forward to 2013, and I'm in awe at what I've been able to accomplish with this framework. From all of the activity that I've been a part of over the past few years, I was led here, to share with you what I know, once again contributing to the now known and still thriving Sencha community.

JAY GARCIA

acknowledgments

Anthony DeMoss—From the beginning, you've worked with me to get this manuscript developed for Sencha Touch 1.0 and then worked feverishly to get it migrated to 2.0. I'm happy we got a chance to work on this together.

Mitchell Simoens—Your knowledge of the framework has been ever-evolving and it's been a pleasure to follow your career growth. Being a coauthor of this manuscript is a testament to how much you've achieved over the past few years.

Abe Elias—You have been an amazing friend and mentor to me. Thank you for your continued advice throughout the years and for affording me the opportunities to work closely with the various Sencha teams and gain experiences that help me push my knowledge forward.

Jamie Avins—Thank you for your continued support with the Sencha Touch framework. Having you available to ask questions and bounce ideas off has been immensely helpful and I'm grateful to you for that.

Jacky Nguyen—As part of the Sencha community, we recognize the technical achievements you've made possible with Touch 2.0. You were there for me when I was faced with technical challenges that were beyond my experience.

Ed Spencer—Even though you're no longer with Sencha, your contributions to my early knowledge of Touch haven't been forgotten. Your presence in the Sencha community is missed.

Don Griffin—Thank you for the back-and-forth when it came to Sencha Cmd! Gaining access to this tool in its infancy helped me close the final chapter for this book.

Sebastian Sterling—The publication of this book has taken a lot longer than we anticipated. As my primary developmental editor at Manning, you challenged my writing and helped me bring out the best content.

Frank Pohlmann—Frank, you came in at a busy point in the development of this manuscript and you did a hell of a job getting things right for production. Thank you.

Liz Welch—Thank you for the hard work in getting this manuscript better organized during the copyediting phase. It's been an absolute pleasure working with you.

Jamund Ferguson—Thank you for your careful technical proofread of the final manuscript shortly before it went into production.

Our reviewers—Thanks for reading the manuscript at various stages of its development and for providing invaluable feedback: Àlex Madurell, Darragh Duffy, Doug Warren, Frank Ableson, Grgur Grisogono, J.J. van de Merwe, Loiane Groner, Matt Goldspink, Pawankumar Tripathi, Robi Sen, Stuart Davies, Tony Niemann, and Vincent Winckelmans.

Finally, to my wife, Erika: You've supported me in everything I've aimed for, and I know that you've sacrificed time with me so that I could share my knowledge with others. You always push me to do better and I'm truly grateful to you. Also to my sons, Takeshi and Kenji: Thank you for the adventures you put your mom through while I worked on this book in my basement "man cave." The stories she's told me certainly have been colorful.

JAY GARCIA

about this book

This book is designed to walk you through the flexible and powerful mobile HTML5 framework, Sencha Touch, from explaining the basics all the way to developing and deploying production mobile web applications. After you've read this book, you should be able to develop mobile HTML5 applications for tablets and phones.

Who should read this book

This book is intended for developers who want to create rich mobile HTML5 applications that feel native using Sencha Touch. Although Sencha Touch is highly customizable, this book is targeted at those who primarily do the programming part of specification implementation.

The book assumes you already have a working understanding of how websites interact with web servers. To be most effective in writing robust and responsive applications, you need a solid background in core technologies like HTML, CSS, JavaScript, and JSON. The only thing we cover in detail about these core technologies is in chapter 10, where we discuss prototypal inheritance with JavaScript, a prerequisite to the Sencha Touch class system.

What you'll need

This book has a lot of hands-on examples that we walk you through. To get the most value out of this book, you need the following:

- A web server (Apache HTTPD or Microsoft IIS recommended)
- An intelligent IDE (WebStorm or Aptana recommended)

- Sencha Cmd installed (available at www.sencha.com/products/sencha-cmd/download)

Roadmap

This book is designed to give you a guided tour of Sencha Touch. During this tour, we'll focus on many of the rich features that Sencha Touch provides, including UI widgets, data stores, models, and proxies. This tour consists of 11 chapters.

Chapter 1 is an introductory chapter, focused on getting you familiar with the framework. This chapter takes a top-down view of the framework, and we'll discuss many of the commonly used widgets. We'll also talk about what it's like to think like a mobile developer.

In chapter 2, you'll get your feet wet and learn how the framework is delivered to you and about its contents. In this chapter, you'll develop a simple application.

Chapter 3 details foundational topics, such as the Component model and the component life cycle. We'll also talk about how Sencha Touch allows you to lazy-instantiate components and how components are tracked with a class called `Ext.ComponentManager`.

Chapter 4 provides a deep understanding of how containers play an important role in your applications and how they implement layouts. You'll learn the ins and outs of the various layouts.

Chapter 5 focuses on how components are "docked" within your UI. You'll also learn how toolbars work and how to implement buttons with proper spacing.

Chapter 6 addresses how to get the user's attention with core widgets such as the sheet, picker, and message box. What you learn in this chapter will come in handy when it comes to notifying a user of activity or requesting information from them.

Chapter 7 gives you an opportunity to take a small break from the UI world for a bit and focuses primarily on data management. You'll take an in-depth look at data stores and their supporting classes, `Model` and `Proxy`. You'll also use data stores with the List and NestedList widgets.

Chapter 8 shows you how to get information from the user via Form panels. We'll discuss the various form input fields and what it takes to use them effectively.

In chapter 9 you'll learn how to render Google maps in your applications. You'll also get an opportunity to use the Sencha Touch Media widget to render video and listen to audio via your browser's native HTML5 media tags.

Chapter 10 focuses on object-oriented JavaScript and shows you how to use the Sencha Touch class system to create custom extensions. You'll get hands-on experience creating a custom extension and a plug-in.

Chapter 11 deals with the development of an application. You'll see how to bootstrap an application with Sencha Cmd and then move on to developing an application using Sencha Touch. Once the application is developed, you'll learn how to create testing and production builds for deployment.

Code conventions

All source code in listings or in text is in a `fixed-width font like this` to separate it from ordinary text. We make use of many languages and markups in this book—JavaScript, HTML, CSS, XML, Java, and JSP—but we tried to adopt a consistent approach. Method and function names, properties, XML elements, and attributes in text are presented using this same font.

In many cases, the original source code has been reformatted: we've added line breaks and reworked indentation to accommodate the available page space in the book. In rare cases even this was not enough, and some listings include line-continuation markers. Additionally, many comments have been removed from the listings.

Code annotations accompany many of the listings, highlighting important concepts. In some cases, numbered bullets link to explanations that follow the listing.

Getting the latest examples

The examples in this book are designed to be easy to navigate. Every chapter has its own folder, with each example named in accordance to the listing to which it corresponds.

We'll work to keep the examples up to date as the framework gets upgraded. To get the latest version of the examples, visit https://github.com/ModusCreateOrg/sencha-touch-in-action-examples or the publisher's website at manning.com/SenchaTouchinAction.

Author Online

Purchase of *Sencha Touch in Action* includes free access to a private web forum run by Manning Publications where you can make comments about the book, ask technical questions, and receive help from the authors and from other users. To access the forum and subscribe to it, point your web browser to manning.com/SenchaTouchinAction. This page provides information on how to get on the forum once you are registered, what kind of help is available, and the rules of conduct on the forum.

Manning's commitment to our readers is to provide a venue where a meaningful dialogue between individual readers and between readers and the authors can take place. It's not a commitment to any specific amount of participation on the part of the authors, whose contribution to the book's forum remains voluntary (and unpaid). We suggest you try asking the authors some challenging questions, lest their interest stray!

The Author Online forum and the archives of previous discussions will be accessible from the publisher's website as long as the book is in print.

Two other author resources are: Jay's blog at http://moduscreate.com/ and Mitchell's blog at http://mitchellsimoens.com/.

about the authors

JAY GARCIA is CTO and cofounder of Modus Create, a company focused on delivering high-end solutions with Sencha products. Jay's involvement with the world of Sencha started in 2006. Since that time, Jay has been focused on knowledge sharing through books, blog articles, screencasts, meetups, and conferences.

ANTHONY DEMOSS is an entrepreneur who has been involved with Sencha-related projects since 2007, focusing on enterprise-level applications. Anthony has developed a custom Sencha Touch 2 front-end for Jira and shares his knowledge via blog posts and meetup presentations.

MITCHELL SIMOENS is the internet forum manager at Sencha. Mitchell's involvement in the community includes helping individuals solve problems and produce professional-grade extensions for Sencha Touch.

about the cover illustration

The figure on the cover of *Sencha Touch in Action* is captioned an "Inhabitant of Novigrad in the lowlands, Dalmatia." Novigrad is a little historic fishing village situated on a narrow inlet of the Novigrad Sea on the Adriatic coast in the region of Dalmatia in what is now Croatia. This illustration is taken from a recent reprint of Balthasar Hacquet's *Images and Descriptions of Southwestern and Eastern Wenda, Illyrians, and Slavs* published by the Ethnographic Museum in Split, Croatia, in 2008. Hacquet (1739–1815) was an Austrian physician and scientist who spent many years studying the botany, geology, and ethnography of many parts of the Austrian Empire, as well as the Veneto, the Julian Alps, and the western Balkans, inhabited in the past by peoples of many different tribes and nationalities. Hand-drawn illustrations accompany the many scientific papers and books that Hacquet published.

The rich diversity of the drawings in Hacquet's publications speaks vividly of the uniqueness and individuality of Alpine and Balkan regions just 200 years ago. This was a time when the dress codes of two villages separated by a few miles identified people uniquely as belonging to one or the other, and when members of an ethnic tribe, social class, or trade could be easily distinguished by what they were wearing. Dress codes have changed since then and the diversity by region, so rich at the time, has faded away. It is now often hard to tell the inhabitant of one continent from another and the residents of the picturesque towns and villages on the Adriatic coast are not readily distinguishable from people who live in other parts of the world.

At a time when it is hard to tell one computer book from another, we at Manning celebrate the inventiveness, the initiative, and the fun of the computer business with book covers based on costumes from two centuries ago brought back to life by illustrations such as this one.

Part 1

Introduction to Sencha Touch

Welcome to *Sencha Touch in Action,* an in-depth guide into the world of Sencha Touch. In this book, in addition to learning how to get things done using the Sencha Touch framework, you're going to learn how the various components and widgets that comprise the framework work and operate.

Chapters 1 through 3 are designed to give you the essential knowledge to understand many of the fundamental aspects of the framework. Your journey begins with chapter 1, where you'll learn the basics of the framework. Chapter 2 is your "boot camp" chapter and is where you'll get your first exposure to developing with the framework. Chapter 3 covers some of the internal machinery of the framework, such as the Component and Container models.

After reading the chapters in part 1 you'll be ready to explore the many widgets that compose Sencha Touch.

1 Introducing Sencha Touch

This chapter covers

- Solving problems with Sencha Touch
- Using the Sencha Touch UI palette
- Thinking like a mobile-web developer

You're on the hook to build a mobile application. Perhaps you've been tasked with a project, or you have a great idea and want to make it a reality. Either way, to build your application you're going to have to learn at least Objective C for iOS or Java for Android. It should be no surprise that if you want to support both types of devices you'll have to learn and master both languages, unless you choose a third-party native framework like Sencha Touch to bridge the gap between the devices.

Chances are you have experience in HTML, CSS, and JavaScript and want to use what you already know to build your mobile application. The ability to tie in your prior experience is part of what makes Sencha Touch a good choice for folks like you and me, because it offers a wide range of UI widgets to choose from, as well as robust data, layout, and component models.

In this chapter you'll begin your journey into the world of Sencha Touch, where you'll learn what Sencha Touch is and the problems it aims to solve, such as enabling

Figure 1.1 Checkout is a full-screen Sencha Touch application that allows you to keep tabs on your GitHub account and followers. You can learn more about it at http://checkout.github.com.

development of cross-platform user interfaces with HTML5. Then we'll look at the widgets that the framework provides. Lastly we'll discuss some of the ways you should think about developing your mobile application to avoid future performance issues.

What you'll learn along the way is that developing mobile applications with this framework isn't as difficult as with other technologies such as Objective C or Java.

1.1 What is Sencha Touch?

Sencha Touch was born out of the culture and many of the ideas from the venerable Ext JS framework and is the first mobile HTML5 JavaScript framework.

Sencha Touch solves cross-platform mobile app development problems by giving developers the tools necessary to build cross-platform applications that mimic natively compiled applications, while making full use of HTML5 and CSS3. It also allows developers who have years of experience on the web to develop cross-platform mobile apps that can exist solely on the web, or be deployed in an app store with either the Sencha native packager or tools like PhoneGap.

As of this writing, Sencha Touch runs on mobile WebKit-based browsers in iOS (iPhone, iPad) devices as well as on Android phones and tablets.

Figure 1.2 The Sencha Touch App Gallery

> **Read about HTML5**
>
> HTML5 is a collection of technologies that includes enhancements to HTML itself, CSS3, and even JavaScript. It's changing the way we develop web applications by providing JavaScript API access to do things like talk directly to a graphics card (WebGL), manipulate sound, and even provide offline storage. Though it's not completely necessary to know everything about HTML5 to use Sencha Touch, it's a good idea to get the basics down. A great site for learning about HTML5 is www.html5rocks.com.

An excellent example of a Sencha Touch application is Checkout, by Steffen Hiller, shown in figure 1.1 running on an iPad. Here you can see an application that makes use of Sencha Touch providing a rich UI with HTML5.

To see other Sencha Touch applications you can point your browser to http://sencha.com/apps and view the Sencha Touch App Gallery (figure 1.2). Here you can preview apps via images and even see them work live via embedded links.

Much like Ext JS, Sencha Touch creates the feel of a native application by means of a clever blend of HTML5, CSS3, and JavaScript, all optimized for the best possible mobile experience given the constraints of mobile devices today, such as limited CPU and memory for your applications.

Also like its big brother Ext JS, Sencha Touch is designed to be extensible and modifiable out of the proverbial box.

1.1.1 What Sencha Touch is not

Although Sencha Touch works on desktop WebKit browsers, like Safari and Chrome (to a limited extent), it isn't designed for desktop rich internet applications. Upon its release lots of developers balked at the idea of this framework not functioning in Firefox or Internet Explorer.

The fact is that Sencha Touch is aimed at the development of mobile applications only. This means that if you've only developed applications with Firefox, IE, and their respective debugging toolkits, you're going to have to leave your comfort zone.

> **"Sencha Touch" !== "Ext JS"**
>
> If you're a veteran Ext JS developer you'll feel right at home when learning Sencha Touch. It's important to know that some significant differences exist between the two libraries. Throughout this book we'll point out some of the differences, but we can't cover every possible point. If you have doubts always check the API documentation.
>
> If you're unfamiliar with Ext JS and need to develop applications for the desktop web check out *Ext JS in Action* (Manning, 2010).

1.1.2 Lots of wiring under the hood

To make use of mobile device interactions Sencha Touch comes with a gesture library, which allows you to easily hook into gesture-based events, such as tap, pinch, and swipe. One way Sencha Touch comes close to the feel of native applications is by means of a custom physics-based `Scroller` class, which uses hardware-accelerated CSS3 transitions and includes key variables like slide friction and spring effects.

1.1.3 Hardware compatibility

Many mobile touch-screen smart devices are entering the marketplace today, which is driving the increase in demand for mobile applications. Though Sencha Touch aims at 100% compatibility across all mobile devices the best user experience is on iOS and high-powered Android devices.

> **Why the difference in user experience?**
>
> The main reason for the difference in user experience between iOS and Android has to do with the physical computing power of each device and how each device manufacturer compiles mobile WebKit for their device. Apple devices include a GPU and compile mobile Safari with GPU acceleration enabled. Most Android devices don't have dedicated GPUs. And even for the ones that do, manufacturers typically don't compile mobile WebKit to enable GPU acceleration.

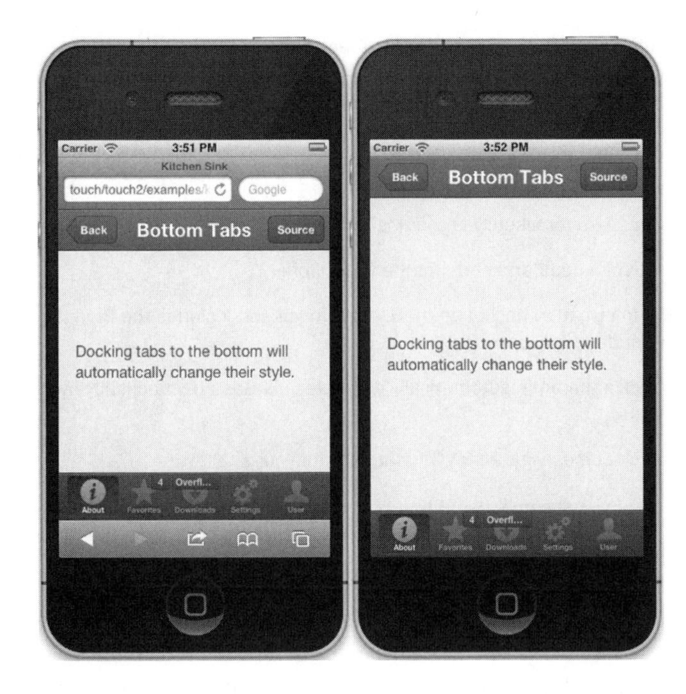

Figure 1.3 Looking at the Sencha Touch kitchen sink example via Mobile Safari directly (left), or a shortcut on your home screen (right)

Sencha Touch applications do such a great job of mimicking how native applications look and feel that it's often easy to get lost in the fact that you're using a web-based application. This especially holds true when the mobile WebKit toolbars are hidden from view.

1.1.4 Full-screen goodness

Figure 1.3 illustrates how a Sencha Touch application looks when accessing the application via mobile Safari (left) compared to accessing the application via a shortcut on your home screen.

After you look at figure 1.3 it should be clear that a full-screen view of a Sencha Touch application closely resembles a native application. Also having your app in full screen means that there's more much-needed screen real estate available in your applications.

Sencha Touch offers a lot when it comes to UI widgets, but it's certainly just the surface of this framework. If you're like us you probably want to just skip ahead and dive into code. But before you get your hands dirty let's browse through the library and discuss some of its features.

1.2 A 10,000-foot view

If you glanced at the Sencha Touch API documentation the sheer number of classes in the library might have overwhelmed you. To make sense of it all you must understand that these classes can be broken down into a few major groups. Table 1.1 describes the major groups into which Sencha Touch is broken down.

Table 1.1 Describing the various sections of the Sencha Touch framework

Group	Purpose
Platform	This is the shared base of Sencha Touch and Ext JS v4, and the bulk of the code for Sencha Touch.
Layout	A set of managers for visually organizing widgets on screen.
Utilities	A group of useful odds and ends for the framework.
Data	Data is the information backbone for Sencha Touch and includes the means for retrieving, reading, and storing data.
Style	Sencha Touch's theme is automatically generated via Sass (Syntactically Awesome Style Sheets).
MVC	Sencha Touch comes with an MVC framework for your application.
UI Widgets	A collection of visual components that your users will interact with.

The base library for Sencha Touch is known as Sencha Platform. Sencha Platform is based off Ext JS 4.0 but is much improved in many respects. The class system in Sencha Touch resembles that of Ext JS 4, but it's much more advanced in many ways. For example, it includes a feature known as Config System which allows Sencha Touch to work much more effectively than Ext JS.

The Layout portion of Sencha Touch is implemented by some of the UI widgets and is the code responsible for visually organizing items on the screen. The layouts are responsible for implementing transitional animations if configured to do so.

The Utilities section of the framework is a collection of useful bits of functionality that are often implemented by the framework and can be implemented by you. For instance, the List widget implements XTemplate to paint HTML fragments on screen. The XTemplate is open for you to use to do the same in your own custom widget.

The Data package is a group of classes that gives Sencha Touch the ability to fetch and read data from a myriad of sources, including mobile WebKit's HTML5 Session, Local, and Database Storage methods. Sencha Touch can read data in a variety of formats, including XML, Array, JSON, and Tree (nested).

The Style area of the framework isn't something that you typically deal with on a day-to-day basis, but it's worth mentioning. From the very beginning Sencha Touch has implemented Sass to allow easy style changes to the UI. This means that if you want to change your entire color scheme you can do so with relative ease if you know Sass.

Learn more about Sass

Sass has taken the world of style sheet management by storm and has arguably revolutionized how people style their web pages and apps. To learn more about this utility check out *Sass and Compass in Action* (Manning, 2013).

Sencha Touch includes an MVC framework that allows developers familiar with that pattern to develop applications within a familiar workspace. It also contains a custom URL routing mechanism and history state support.

The widgets that users will interact with in your application comprise the UI portion of Sencha Touch. When thinking about designing and constructing your applications you have a lot to choose from, which is why it's a good idea to look at each of them.

1.3 The Sencha Touch UI

The Sencha Touch UI is a rich mixture of widgets that can be displayed on screen for you and your users to interact with. The UI palette is large, and table 1.2 helps you identify the groups of UI components.

After reviewing the groups we'll dive deeper into each group and discuss the UI components in greater detail.

Table 1.2 The various groups of UI widgets available in the Sencha Touch framework

Group	Purpose
Containers	Widgets that are designed for nothing more than managing other child items. An example of these types of widgets is the Tab panel. Containers typically implement layouts.
Sheets	Sheets are generally any popup or side-anchored container and appear in a modal fashion, requiring users to interact with the sheet before moving forward. An example of a sheet is the Date Picker widget.
Views	Views are widgets that implement data stores to display data. The List and Nested List are both views that implement Stores.
Misc	This collection of widgets ranges from Buttons to Maps to Media.

Now that you have a good overview of the widget groups let's begin our visual exploration.

1.3.1 Containers

Containers in Sencha Touch do the heavy lifting when it comes to managing widgets inside of widgets. *Container* is what we like to call the workhorse of Sencha Touch applications because it offers extreme configurability and flexibility. Containers can dock child widgets to their sides or render child widgets inside of their bodies. To see what we mean take a look at figure 1.4.

In figure 1.4 you see a container with three docked items. We have a toolbar docked at the top, List view docked on the left, and another toolbar docked at the bottom. Notice how the top-docked toolbar simply contains a title, whereas the bottom toolbar contains buttons. This shows some of the power and flexibility of the toolbars.

If you need to display screens controlled by a toolbar the Tab panel will get the job done.

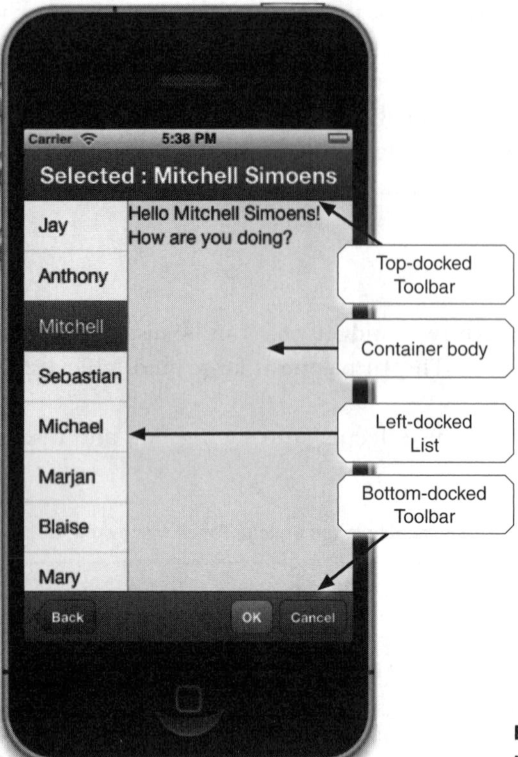

Figure 1.4 A demonstration of some of the panel content areas

1.3.2 Controlling your UI with the Tab panel

The Tab panel (figure 1.5) is a container that automatically sets a top-docked or bottom-docked toolbar for you with automatically generated buttons for every child item. Tapping any of the buttons allows you to "flip" through items known as "cards."

In figure 1.5 we've configured a Tab panel that implements a "slide" transition with two child panels. By selecting Panel 2 in the toolbar Sencha Touch automatically applies the CSS3 transition properties to both child panel elements, allowing for a smooth transition from one panel to another.

The Tab panel does an excellent job of managing the display of items in your screen, but sooner or later you'll need the ability to accept data input from your users. For that you'll use the Form panel.

1.3.3 Accepting input with the Form panel

The Form panel is a container that's typically used to display any of the input fields that Sencha Touch provides and is automatically scrollable. Your fields can be grouped via the FieldSet widget. Figure 1.6 shows an example of a Form panel requiring user input.

In figure 1.6 we have most of the input fields that Sencha Touch offers, with the exception of the Hidden field. With Sencha Touch the Text, Checkbox, URL, Email,

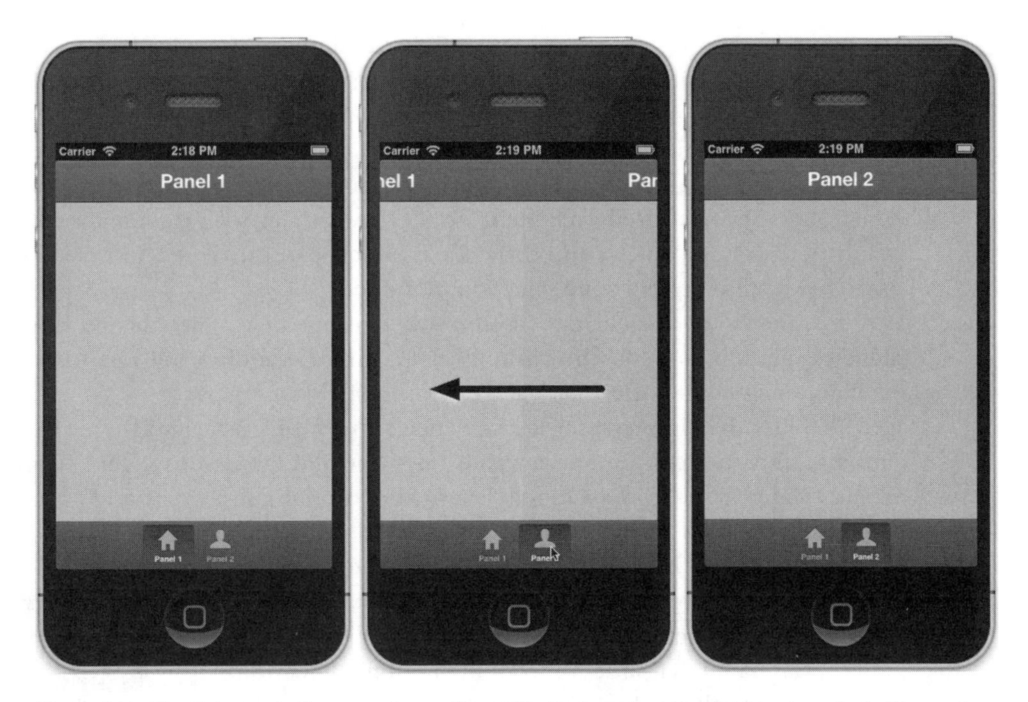

Figure 1.5 The Tab panel allows you to configure UIs that can be changed by a tap of a button and includes optional transition animations (from left to right).

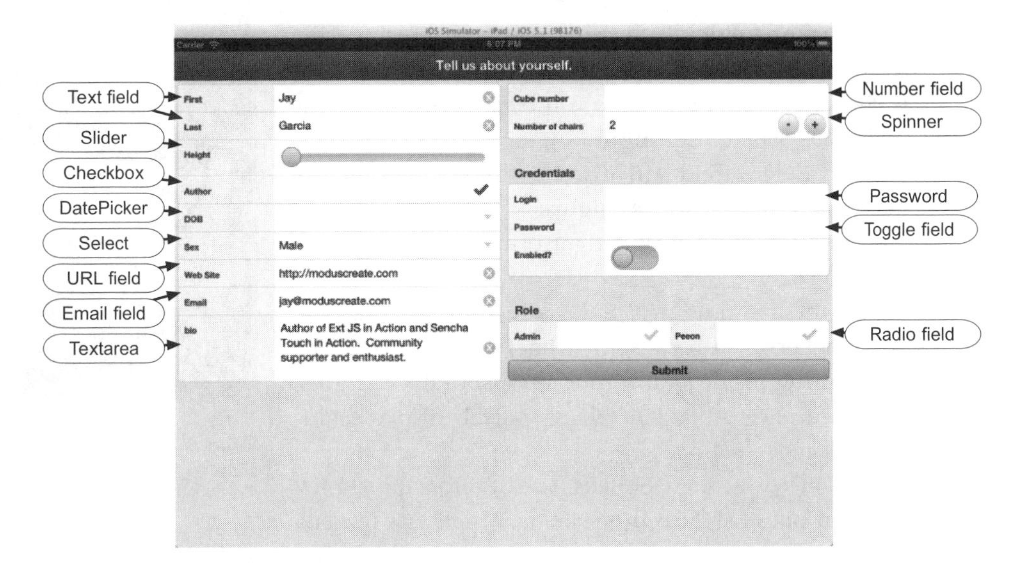

Figure 1.6 Form panels are used to display input fields and contain necessary controls to manage the submission of data to your server.

Textarea, Number, Password, and Radio fields all implement native HTML5 input elements, with the addition of styling. Each of these, except for the Radio and Checkbox fields, will force the native slide-in keyboard to appear when focused, allowing users to enter data into the fields.

The Checkbox and Radio fields work similarly to their native web counterparts, except that they're stylized via Sencha Touch's own check icon to mimic native application behavior. In this example the Role checkboxes, grouped in a Fieldset, are Radio fields, allowing only one selection in the set.

Next the Spinner field is a custom-styled input field, allowing users to enter numeric values, much like the Number field, with the addition of easy-to-use decrement and increment buttons on each side of the field.

The Slider field implements native Sencha Touch Draggable and Droppable classes, allowing users to input a numeric value, via swipe and tap gestures. The Toggle field extends Slider, allowing users to toggle a field of two values via swipe and tap gestures, much like an on/off toggle switch that you see in various physical devices.

Lastly, the Date Picker and Select fields give your users the ability to choose data from a set. The Date Picker field implements what's known as a sheet, which is an overlay panel that slides in from the bottom allowing the user to select values via vertical swipe or "flick" gestures. Figure 1.7 shows an example of a date picker displayed on an iPhone.

No matter what the device or its orientation the Date Picker field will always display a sheet forcing selection through this modal overlay. The Date Picker widget may seem familiar to you because it mimics the native iOS Date Picker input widget.

The Select field will display different input widgets based on the device. Figure 1.8 shows our implementation of the Select widget on a phone and on a tablet device.

On the left in figure 1.8 the Select widget is displaying a Picker sheet because Sencha Touch detected that it's running inside of a phone versus a tablet (right). The difference is that iOS tablets natively display dialog controls for selection.

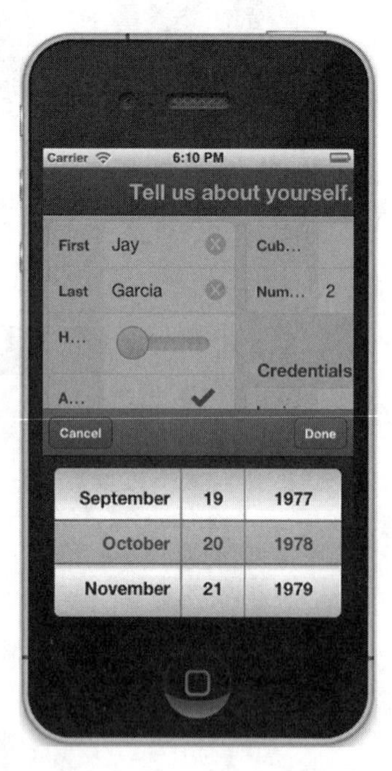

Figure 1.7 The Date picker slider in action

As you've just seen, Sencha Touch offers quite a few wrapped native HTML5 input fields as well as a few custom widgets. Because we've been talking about the Date Picker and Select fields implementing sheets let's look at the various Pickers and Sheets that Sencha Touch offers, outside of the Form panel.

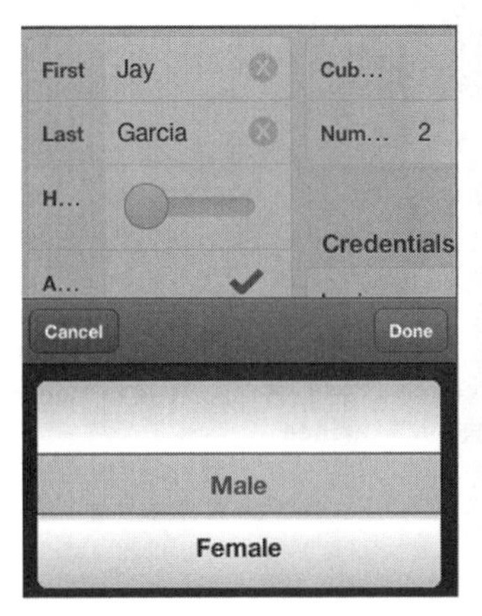

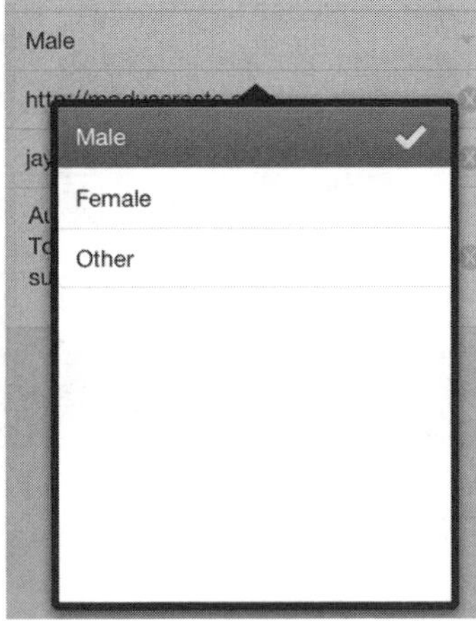

Figure 1.8 The Select field will display a different input widget based on the type of device that's running your UI.

1.3.4 *Sheets and pickers*

We've already seen the `Date picker` and `Picker` classes implemented via their associated form input widgets. Sencha Touch provides you with a widget called a sheet, which is a floating modal panel that animates into view, grabbing the user's attention and focus. Figure 1.9 shows a sheet in action.

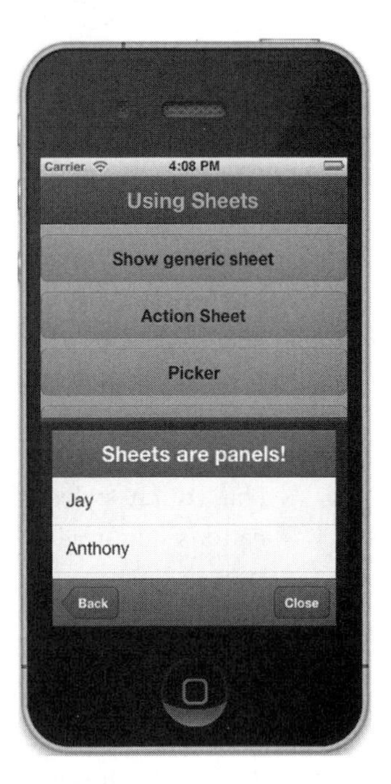

Figure 1.9 A generic sheet in action displayed in portrait mode

In figure 1.9 you can see a sheet with top- and bottom-docked toolbars, managing a scrollable List view. You can configure such a UI because `Sheet` is a subclass of `Panel` and includes all of the UI goodness that `Panel` provides.

What's neat about the Sheet widget is that it's orientation-aware. This means that flipping the device while the application code is executing immediately causes the sheet to render in landscape mode, as illustrated in figure 1.10.

The story about `Sheet` doesn't end here. It has three subclasses: `ActionSheet`, `MessageBox`, and `Picker`. You've already seen `Picker` and its subclass, `Date picker`, but you haven't seen `ActionSheet` and `MessageBox`.

We're throwing a lot of new names at you, so to help with any confusion see the simple inheritance model diagram (figure 1.11).

Out of the box the ActionSheet widget allows you to easily render buttons in a sheet, rendered in a vertical stack. Because `ActionSheet` is a subclass of `Sheet`, which extends `Panel`, you can add pretty much anything you want to the stack of widgets. Figure 1.12 shows an example of an ActionSheet widget rendered with a custom HTML title.

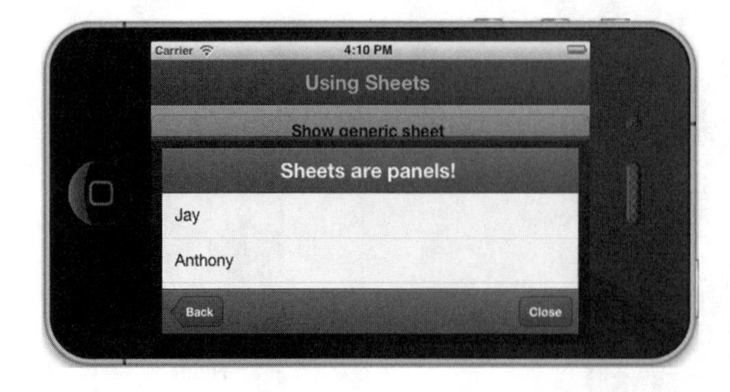

Figure 1.10 A generic Sheet widget displayed in landscape mode

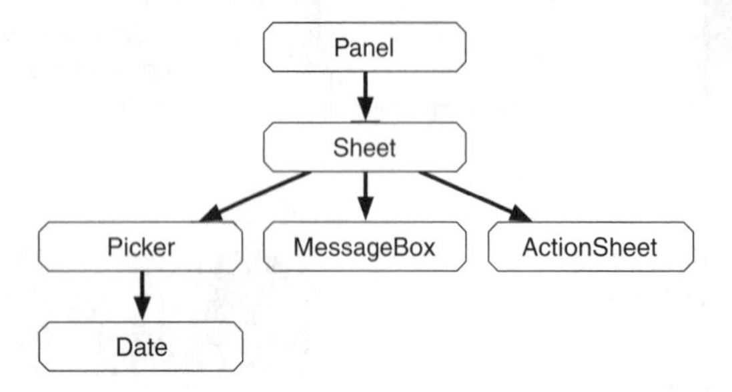

Figure 1.11 The Sheet inheritance model

Such an ActionSheet could be used to request action from users, requiring that they choose an action via one of the buttons. In this case you're asking the user to choose one of three options, and with the custom title you're providing a hint along with the actionable button set.

`MessageBox` is a subclass of `Sheet` that provides Sencha Touch–styled alert-like functionality to your applications. Figure 1.13 illustrates the three most common uses of `MessageBox`, including alert (top left), confirm (bottom left), and prompt (top right). Each of these dialogs appears with smooth CSS3 transitions and mimics their native counterparts.

The key differences among the three are apparent. The alert MessageBox widget is designed to alert the user of some condition and only displays one button. The confirm dialog allows the user to make a decision by tapping on a button, enabling a branch of logic to execute. Lastly, the MessageBox prompt asks the user for direct input.

You've seen all that `Sheet` and its subclasses have to offer. Next let's look at the various data-bound views that Sencha Touch provides.

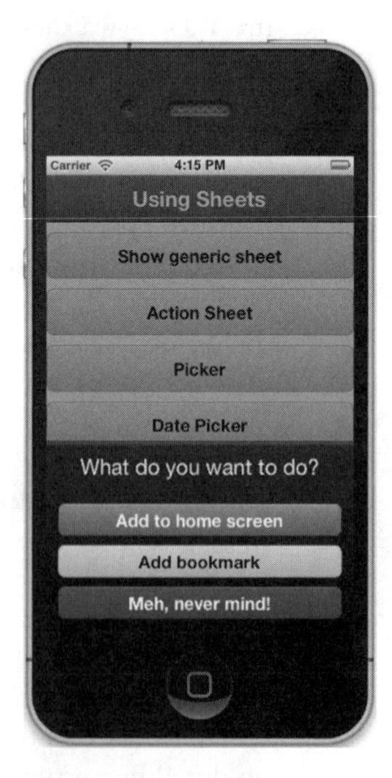

Figure 1.12 An ActionSheet displayed in an application

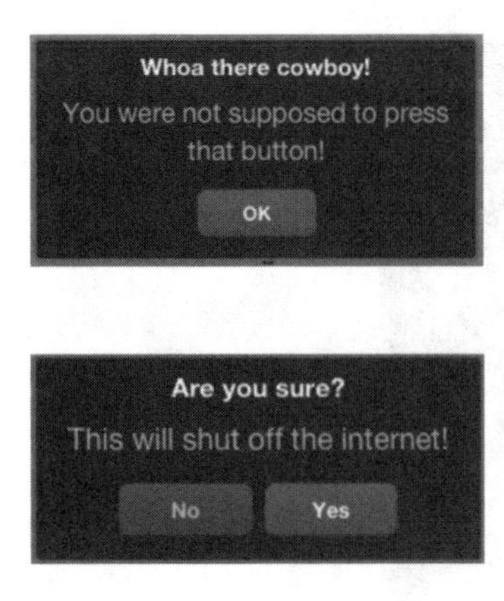

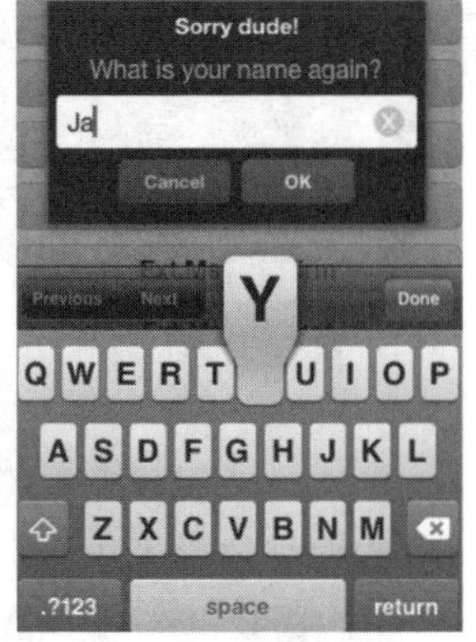

Figure 1.13 The common MessageBox implementations: alert (top left), confirm (bottom left), and prompt (right)

1.3.5 *Data-bound views*

If you're an Ext JS developer you might be surprised to learn that Sencha Touch only provides three data-bound views and that this list excludes a GridPanel. At your disposal you have Data View, List, and NestedList. We'll begin with the most basic, Data View, and work our way to the most complex.

Data View is a widget that binds to a data store to render data on screen. It gives you 100% control over how you'll render your data. Figure 1.14 is an example of a simple Data View widget displaying a set of names, beginning with the last name.

Here we have a stylized Data View widget rendering data from a data store, which contains a list of names. This example rendered only names to keep it simple but you can use Data Views to render anything imaginable and to allow for user interaction.

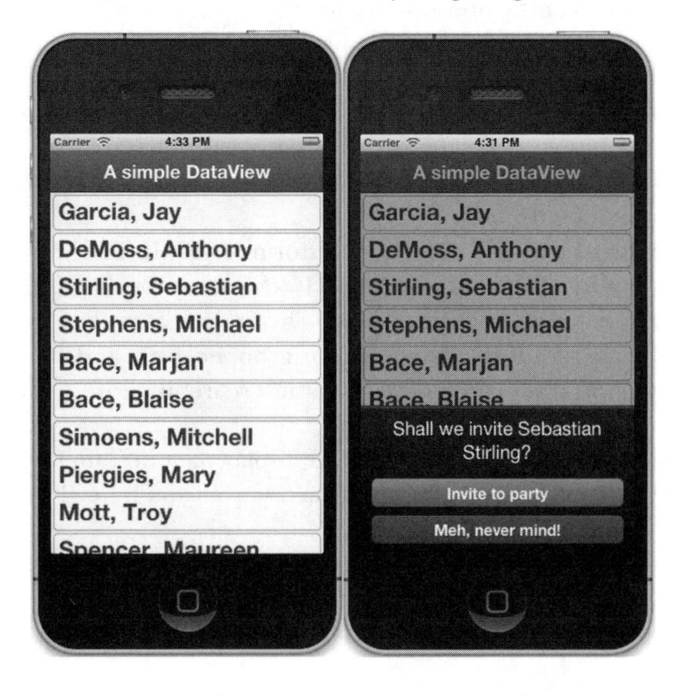

Figure 1.14 A stylized Data View widget with an "itemtap" event handler, displaying an action sheet based on selection

Figure 1.15 A simple List View with an action sheet

With the Data View widget you're completely responsible for a lot of work, including defining the XTemplate that'll be used to stamp out the HTML fragments, as well as styling how the items are rendered on screen. If you're like us and want the look and feel of a native list the List class is at your disposal. Figure 1.15 shows a List widget, rendering the exact same data as in the previous example.

The difference between this example and the previous one should be clear. What's not as obvious is that the level of effort required to create this List view is orders of magnitude less than creating the earlier Data view. You'll see what we mean by this in chapter 6, when we tackle List views head on.

In addition to allowing for a native application look and feel the List view has three more key features, shown in figure 1.16.

With a few minor tweaks you're able to transform a simple List view into a grouped List view. The grouped list in figure 1.16 has what's known as a grouping bar, which is a separator between items in the list. The Sencha team has been able to get this list to work nearly identically to native grouped lists, and it includes optional disclosure icons, as well as an index bar for fast searching with a single finger swipe.

The Data View and List widgets are designed to display data in a linear set. But there are times when you want to display nested data. For that you'll need to use the NestedList widget, shown in figure 1.17.

In figure 1.17 we've set up a NestedList for the selection of a food item. There are two main categories: Drinks and Food (left). We chose "Drinks," which brought us to three subcategories (center). We then chose "Sports Drinks," which led us to the last

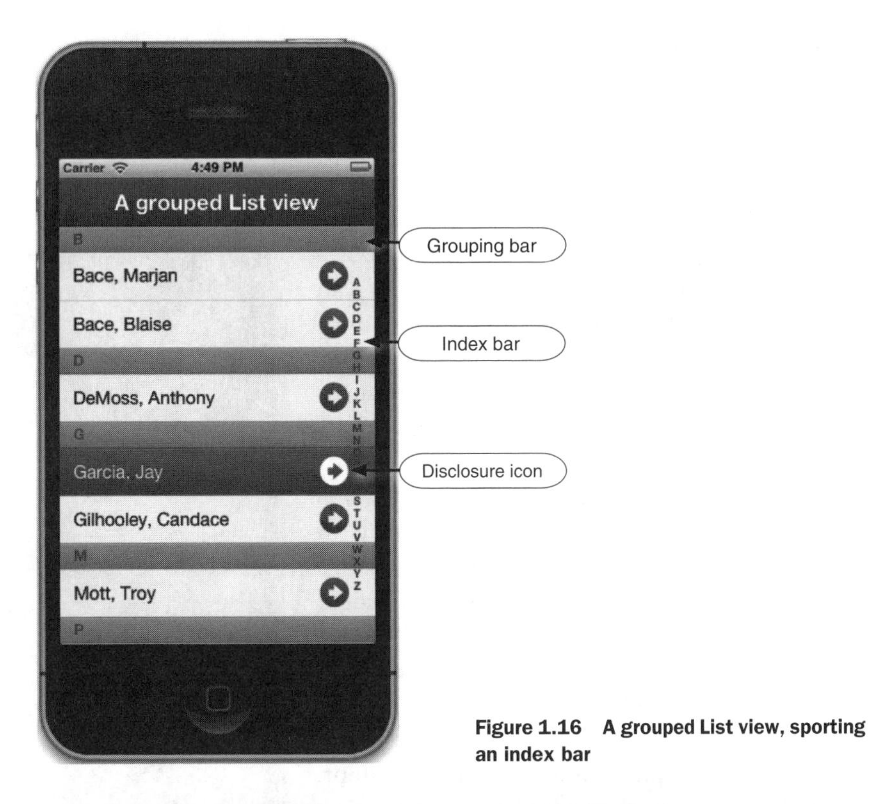

Figure 1.16 A grouped List view, sporting an index bar

Figure 1.17 A NestedList widget in action used in a navigational manner

section of items, which is a list of specific sports drinks (right). All of this is done with the slide animation.

You've just explored the Data View, List, and NestedList widgets. Next let's examine the Map and Media widgets.

1.3.6 Maps and Media

With the rapid-expansion world of mobile applications integrating maps into your applications can provide a huge boost in productivity for your users. To meet this growing demand Sencha Touch integrates Google Maps to supercharge your location-aware applications. Figure 1.18 shows the Sencha Touch Map widget in action.

Figure 1.18 The Sencha Touch Map widget in portrait mode

The Sencha Touch Map widget literally *wraps* Google Maps, allowing your application code to manage the Google Maps instance as if it was native to Sencha Touch. This means that the Map widget can take part in layouts and has normalized events, as well as an interface method to easily update the map's coordinates.

Another growing demand for mobile applications is the ability to play audio and video content. HTML5 natively has video and audio tags that bring this functionality to Mobile Safari but Sencha Touch makes it easier to use. Figure 1.19 shows the Media widget displaying a video on a tablet.

Just like the Map widget, the Media widget uses familiar interface methods, and it's easily configurable to play audio and video in your applications.

We've just completed our UI walkthrough. Before we wrap up this chapter we want to talk about thinking like a mobile developer. This conversation will be especially helpful to you if this is your first dive into the world of mobile application development. We know you're itching to get down to coding so we won't hold you very long.

1.4 Thinking like a mobile developer

If you're like us you're making the transition from Ext JS to Sencha Touch. Making the transition to mobile from desktop application development poses thought-process challenges that must be overcome if you plan to build successful apps. Here are some points you need to think about before moving forward with your application development.

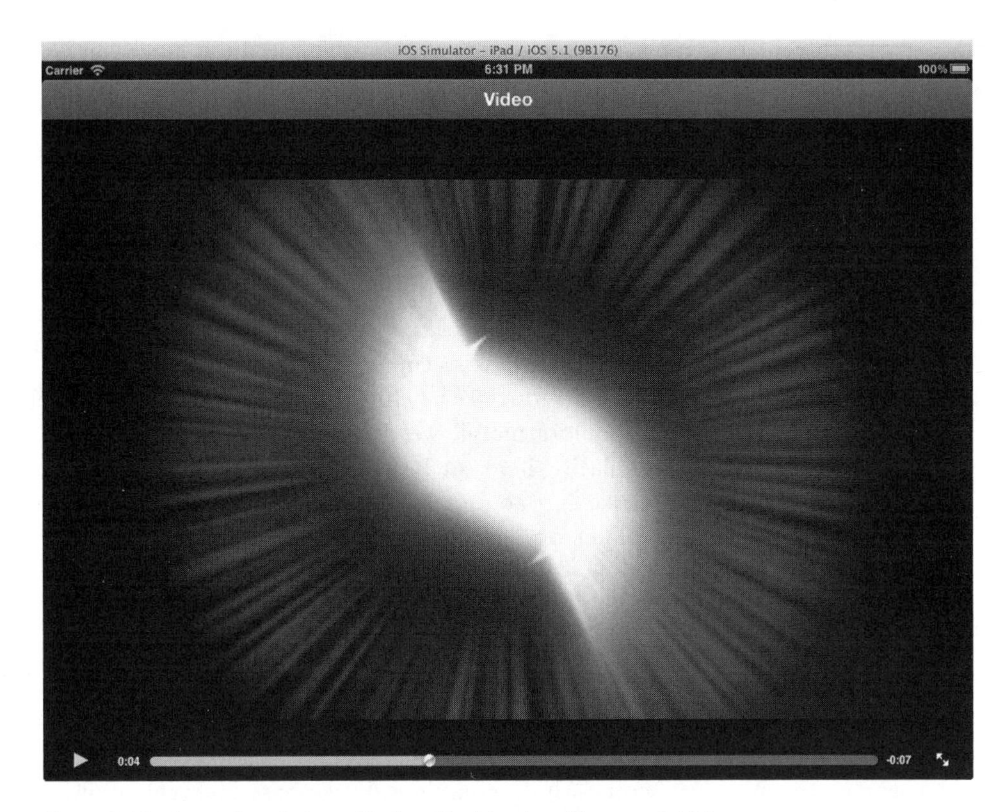

Figure 1.19 A panel wrapping a Media widget to play video on a tablet

1.4.1 *Think lightweight*

When spec'ing out or developing your mobile app you must think "lightweight" or your app is destined to run into performance issues. If you've made the transformation from a native desktop application developer to a desktop-web application developer it's likely that you've encountered this issue during your transition, because native desktop applications can handle much more of a burden than desktop-web applications.

Due to the reduced computing power of mobile devices the mobile browser is limited in many ways when compared to its desktop counterpart. This is why thinking "lightweight" is paramount for a successful application.

Our suggestion is to try to reduce the amount of data as well as the complexity of the screen size as much as possible. Reducing the user interaction models is also a plus since complex user interaction models bog down mobile-web applications.

1.4.2 *Remember—it's a browser!*

Many developers are tasked with converting native applications to Sencha Touch–powered web applications. Often during the conversion process they experience performance issues, and Sencha Touch is blamed.

It's during these times that we tell developers caught in this cycle to remember that the application they're developing *is* running inside of a browser and thus has limited power relative to native-compiled mobile applications. Just as native desktop applications can handle more difficult tasks than desktop-web applications, native mobile apps have more muscle than mobile-web applications.

We believe that entering the conversion process with this in mind helps you set realistic expectations with your customers.

1.4.3 *Throw away what you don't need*

With the reduced power of mobile devices comes an increase of responsibility to keep things as clean as possible and reduce DOM clutter and bloat. For desktop-web applications this isn't as critical, but for mobile web it's extremely critical.

This means that when placing items such as ActionSheets in the Document Object Model (DOM) you must take care to destroy them when they are no longer needed. DOM bloat is the enemy of performance.

> **Mobile Safari will crash**
> Mobile Safari will crash if your application causes it to run out of memory. This will simply cause the application to disappear from the user without warning.

Sencha Touch widgets come with a complete three-phase lifecycle, allowing you to easily destroy components, thus removing items from the DOM and freeing up crucial resources.

Along with the destruction of items that are no longer needed you should only instantiate what's needed. We often find hugely nested components being instantiated when only a single component is needed for a particular action. To keep things safe think conservatively.

1.4.4 *"finger" !== "mouse"*

Part of transitioning to mobile development involves understanding the user interaction models and how they relate to browser events. Table 1.3 describes some of the most common user gestures, alongside their desktop counterparts (when applicable).

Table 1.3 Comparing touch gesture events with desktop mouse gesture events

Mobile	Desktop	Description
touchstart	mousedown	The initial point at which a touch is detected in the UI.
touchend	mouseup	Signals the end of a touchstart event.
tap	click	A tapstart and tapend event for a single target.

Table 1.3 Comparing touch gesture events with desktop mouse gesture events *(continued)*

Mobile	Desktop	Description
doubletap	doubleclick	Two successive tapstart and tapends for a single target.
swipe		A tapstart and tapend event with a delta in X (horizontal) or Y (vertical) coordinates.
pinch		A complex multifinger "pinch" gesture. It consists of multiple touchstart and touchend events with deltas in the X and Y coordinate space.

Always test all of your complex interaction models with the physical platforms that you're targeting for your applications. It's only then that you can truly see how they'll react during events like pinch, swipe, and drag.

1.4.5 Reduce the data

When developing your applications you have to remember to reduce the amount of data you're sending to the browser. If you find yourself pushing megabytes of data to the server for a single Ajax request reconsider your approach.

Along these lines, also aim for a reduction in data complexity. Remember that these devices are relatively low-powered and any time spent manipulating complex data could be spent allowing the user to interact with the application. Tasking your mobile application to deal with deeply nested and complicated data structures is highly discouraged.

Server-side developers will have to work harder

Often deeply nested data structures are passed to clients because of the amount of work involved for the server-side developers to reduce complexity. We'd much rather have our server-side developers work harder than impact the performance of the client and thus avoid a negative view of our mobile applications.

Through your application development iterations we suggest placing yourself in the shoes of your end user. Remember, mobile applications should be quick and responsive.

1.5 Summary

We covered quite a bit in this first chapter, beginning with a high-level discussion about what Sencha Touch is and the problems that it aims to solve for the mobile-web application space.

You then took a deep dive into the world of the Sencha Touch UI widget set and learned about what's offered out of the box. You explored some of the differences between widgets, such as the Data View and List view.

Finally, you learned how you should think about mobile applications and some of the limitations that mobile devices pose.

In the next chapter we'll begin our deep dive into Sencha Touch, beginning with where to get the framework, and then we'll inspect its contents. After you've become familiar with setting up a basic Sencha Touch app page you'll develop a quick application with the framework.

Using Sencha Touch for the first time

2

This chapter covers

- Downloading Sencha Touch
- Exploring the package contents
- Creating a "Hello World" example
- Developing an application

In the previous chapter you looked at what Sencha Touch provides developers. That chapter effectively set the stage for you to begin using the framework, and that's precisely what you'll be doing next.

In this chapter you'll begin to understand how pieces of Sencha Touch work and fit together as you start to develop an application using critical pieces of the framework such as the class `Loader`. Because this is an introductory chapter we won't be going into too much detail about how things operate. We'll go into more detail starting with chapter 3.

2.1 License considerations

To get a copy of the Sencha Touch framework you'll need to visit http://sencha.com/products/touch/download/. You'll be greeted with a page like the one in figure 2.1.

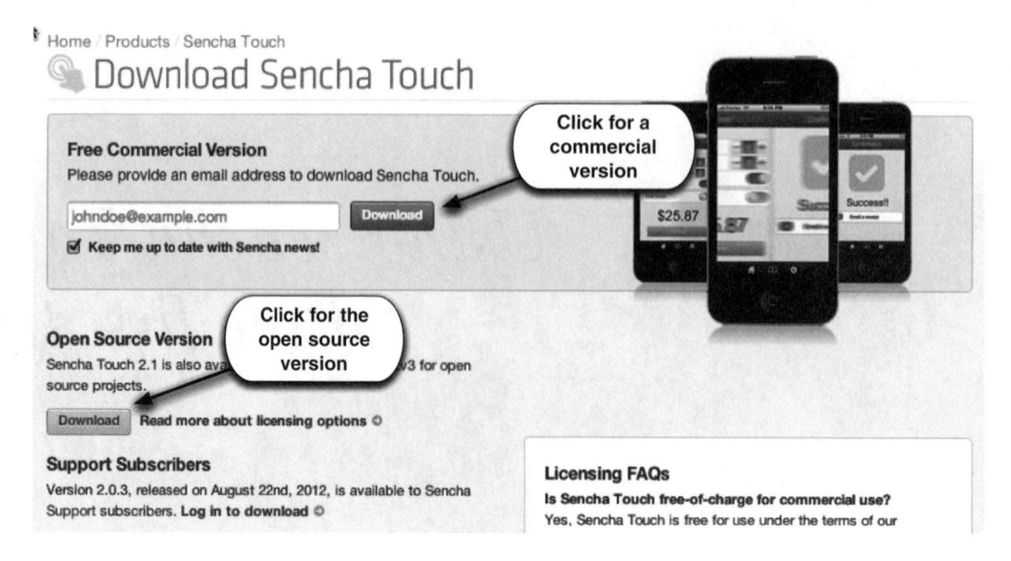

Figure 2.1 You must make a decision about which version of Sencha Touch you're going to download.

From here you're required to make a decision about which license you're going to use. In case you're not sure which version to choose we've detailed a few scenarios in table 2.1.

Table 2.1 Choosing a license model for Sencha Touch

License Type	Reason
Open Source	Your project will be available for anyone to download and modify according to the GNU GPL license v3. FLOSS (Free/Libre and Open Source) exceptions are available.
Commercial	You're creating an application that you plan on selling or distributing and aren't willing to permit direct access to your source. If you plan on packaging your product to offer in an app store this is the license you want to go with.

If you're still unsure which license to choose you can consult the Sencha licensing page at http://sencha.com/products/touch/license or email the Sencha license support team at licensing@sencha.com.

Now that we've gotten the hairy licensing issue out of the way we can begin to look at the package contents of the framework.

2.2 Unpacking the framework

The Sencha Touch framework comes to you via a zip file which you must unpack in order to use the library. Figure 2.2 shows what you can expect to find in the zip file you downloaded.

The zip package inflates to a directory that is just over 270 MB. If you're wondering if all of the package contents are required to deploy with your application, the answer is no.

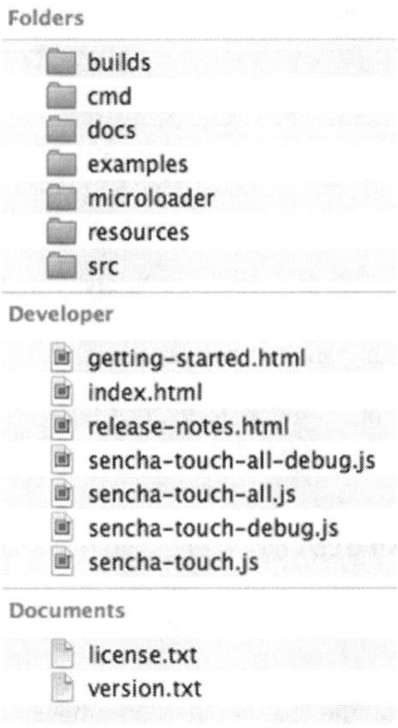

Folders

- builds
- cmd
- docs
- examples
- microloader
- resources
- src

Developer

- getting-started.html
- index.html
- release-notes.html
- sencha-touch-all-debug.js
- sencha-touch-all.js
- sencha-touch-debug.js
- sencha-touch.js

Documents

- license.txt
- version.txt

Figure 2.2 Looking at the package contents

The good news is that most of the space used in the package goes to docs, examples, and other goodies that you typically don't and shouldn't deploy with your applications. See table 2.2 for a description of each of the items in the archive that you just extracted.

> **What you have may look a bit different**
> As of this writing we're covering Sencha Touch 2.2. It's known that the SDK contents change a bit from version to version.

Table 2.2 The various files in the Sencha Touch package

Item(s)	Purpose
builds/	This directory contains the sencha-touch-compat.js file, which contains the code you'd use if you wanted to run your app with backward-compatible 1.0 code. Don't depend on this version of the framework for your application code; we suggest migrating to 2.0 as quickly as possible.
cmd/	This directory contains necessary files for the Sencha Cmd toolset, which is like a Swiss army knife of the Sencha Touch 2 development world. We won't be diving into Cmd until we talk about advanced topics in chapter 11.
docs/	The docs directory is where you'll be able to view the framework documentation without the need to go online. This directory itself is over 139 MB in size.
examples/	This folder contains all of the one-off examples as well as a few example applications. This folder is also responsible for the hefty weight of the unzipped package—it's over 113 MB in size.

Table 2.2 The various files in the Sencha Touch package *(continued)*

Item(s)	Purpose
microloader/	This directory contains various versions of the Sencha microloader code. We'll be covering this in greater detail in Chapter 11, but for now I'll tell you that the microloader is responsible for bootstrapping your application code from start to finish.
resources/	This folder contains all of the CSS and image files required to style the framework. It also has a subdirectory named "sass," which includes all the Sass code needed to get you started developing your own custom themes.
src/	Here's where all of the complete source files are located. You'd typically view the src directory contents if you wanted to take a deep dive into the framework and learn how the widgets work.
getting-started.html and index.html	getting-started.html is a rather lengthy write-up on how to get started with Sencha Touch, and the index.html file serves as a landing page for you if you ever browse the contents of the framework directory. From here you can launch the API documentation and example pages.
release-notes.html	This document describes in great detail all the changes to the framework since its initial release.
sencha-touch*.js	Four "sencha-touch" js files are included, each with varying sizes. We typically use sencha-touch-debug-w-comments.js (not shown in figure 2.2) when debugging locally inside of a browser, and we use sencha-touch-debug.js when debugging on a device. The comments file contains extra weight that makes it 1.54 MB in size, whereas the debug file without comments is just over 750 K. When going to production you should build a custom version of Sencha Touch via Sencha Cmd. This will give you an optimized build based on your application needs.

As you can tell, there's a lot in the Sencha Touch framework out of the box. When it comes to our applications we typically rely on what Sencha Cmd outputs (see chapter 11 for a sneak peek) to generate project structure. For docs and examples reference we refer to a virtual host on our dev machines, which contain each version of the framework for reference.

Now that you have a good understanding of the package contents you can begin to set up Sencha Touch on a dev web server.

2.3 Sencha Touch says "Hello World"

To get this party started you're going to create an HTML file that you'll have to populate with HTML and required JavaScript and CSS references. The following listing shows the test page.

Listing 2.1 The HTML body for your helloworld.html file

```
<!DOCTYPE html>
<html>
<head>
    <meta http-equiv="Content-Type" content="text/html; charset=utf-8">
    <link rel="stylesheet"
href="js/sencha-touch-2.0.0/resources/css/sencha-touch.css"
```

1 Includes Sencha Touch CSS

```
type="text/css">

    <script type="text/javascript"
➥ src="js/sencha-touch-2.2/sencha-touch-debug.js"></script>

    <script type="text/javascript"></script>
</head>
    <body>
    </body>
</html>
```

❷ References debug JavaScript

❸ Includes empty script tag

In listing 2.1 you have all the necessary structured HTML for your simple "Hello World" page. It includes the required tags for the Sencha Touch CSS ❶ and JavaScript ❷ files, as well as an empty script tag ❸ that you can use as a simple playground.

If you load this page now you're only going to see a blank page. That's because you haven't implemented anything from Sencha Touch yet.

The next listing contains code that'll launch the Sencha Touch MessageBox, providing an indication that everything is working well.

Listing 2.2 Using Sencha Touch for the first time

```
<script type="text/javascript">
    Ext.application({
        launch  : function() {
            Ext.Msg.alert(
                'Hello!',
                'Hello there from Sencha Touch!'
            );
        }
    });
</script>
```

❶ Calls Ext.application

❷ Contains Ext.launch function

In listing 2.2 you use Sencha Touch for the first time, rendering the MessageBox on your screen. Here's how all of this works.

To bootstrap your code you execute `Ext.application` ❶ inside your playground JavaScript code block. `Ext.application` is a utility function that's used to configure Sencha Touch for use on your mobile devices and should be called only once, as it's designed to assist with bootstrapping of your code if you don't intend to use the Sencha MVC framework.

You pass one parameter to `Ext.application`, which is known as a configuration object. Within the config object is one property, known as `onReady` ❷, which is executed by Sencha Touch when it detects that WebKit has the document fully loaded and is ready for direct manipulation.

Know thy config object

Because you'll see the term *config object* used a lot in the Sencha Touch API documentation and this book it's important that you know exactly what we're talking about. Like Ext JS, in Sencha Touch most constructors and some methods use config objects, which are used by classes to modify how the class or widget behaves.

Within the `onReady` function is your use of the `MessageBox` class, accessible via the `Ext.Msg` singleton. To get the MessageBox widget to display you call the `alert` method and pass three parameters: `title`, `message body`, and `callback`. The `callback` is a reference to a reusable empty function that resides inside the `Ext` namespace.

> **Ext namespace?**
>
> Sencha Touch uses the `Ext` namespace because it shares a common codebase with its big brother, Ext JS. This common codebase is known as the Sencha Platform. If the Sencha Touch framework code were to be moved to a "Sencha" namespace some unnecessary weight would be added to the framework.

Your helloworld.html file is ready to be viewed in a browser or device. Figure 2.3 illustrates your first Sencha Touch code execution.

Success! Your first Sencha Touch code is live and working. As illustrated in figure 2.3, your code is executable in a desktop browser as well as on a mobile device. In this particular instance your code is running inside Chrome.

Great! Now you've gotten your hands a little dirty with Sencha Touch, but you're not done yet. You did this simple demo so you can ensure that your first Sencha Touch project is configured correctly and functioning properly.

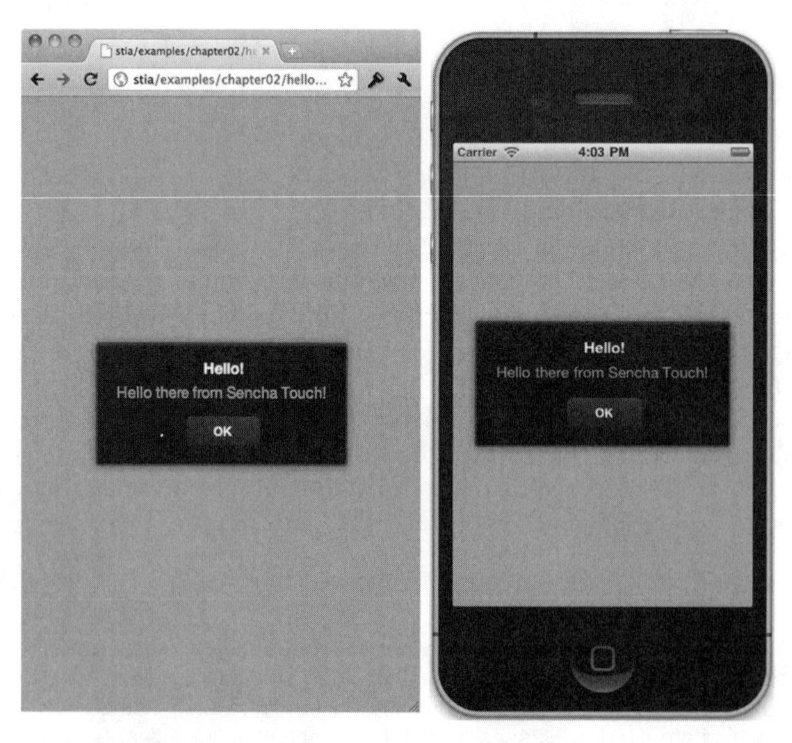

Figure 2.3 Your helloworld.html in a browser and on a phone

> **Your desktop browser choices**
> Because Sencha Touch is designed to work with WebKit browsers only, the choice of desktop-based development and debugging tools is limited to two browsers, Safari and Google Chrome. Both of these browsers are based on the WebKit engine but are somewhat different. As of this writing Safari's 3D CSS is fully supported by Sencha Touch, whereas Chrome's 3D CSS is currently in the experimental phase and not fully functional.

You just learned a bit about `Ext.application`. Doing all of this work sets the stage for you to develop a basic application with this mobile framework.

2.4 Setting the stage for your first application

Because this is most likely your first attempt at creating a Sencha Touch application we're going to take things easy. For instance, you won't be worrying about any target devices, orientation, or even MVC.

Also, you won't implement any server-side code. Instead you'll use an existing API online and fetch data via JSON-P. Before you can write a single line of code you need to know what the application is going to do.

2.4.1 Your simple application at a glance

Today you're going to build a relatively simple contact manager application. This application will allow you to view details of records and will have a left-to-right workflow. Figure 2.4 shows what it'll look like on a tablet.

You can see that you'll be implementing List and FormPanel widgets along with some toolbars. Your app will implement a typical left-to-right workflow, where you'll be able to select a record on the left and see its details on the right in the `FormPanel` instance.

Your next step is to prepare a project that allows you to construct this application.

2.4.2 Preparing your project

To construct the foundation of your code you'll need first to create some files and a directory to organize your class structure. You'll be creating these separate files because we want you to get used to the idea of having one class per file, because it's considered a best practice by the Sencha community and follows object-oriented programming (OOP).

You'll be using the class `Loader` system, so you'll need to create some files and folders in your development environment as shown in figure 2.5. For now, just make these empty files. You'll fill them out along the way.

In our example you have one directory, MyApp, with several files. Inside the MyApp directory are the five JavaScript class files you'll soon create.

Table 2.3 details what function each file will perform.

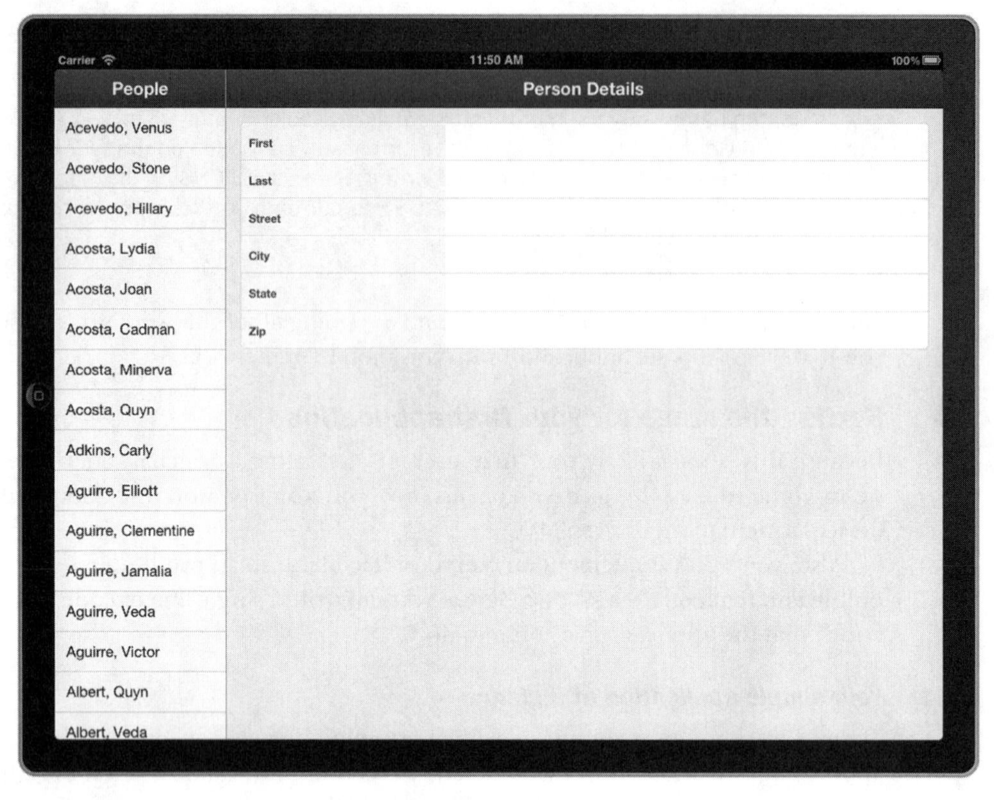

Figure 2.4 A preview of your contact manager app

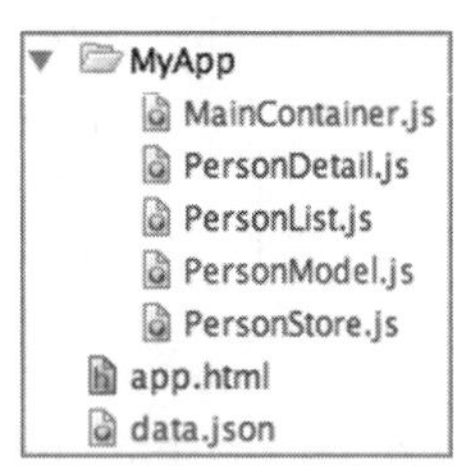

Figure 2.5 The directories that implement the MVC code structure

Table 2.3 The list of files in the MyApp directory and the role of each in your application

File	Role
MainContainer.js	This file extends `Ext.Container` and will be a home for the `PersonDetail` and `PersonList` classes using the VBox layout.
PersonDetail.js	Here you'll extend `Ext.form.Panel` and render a few text fields, displaying the details of the `PersonModel` instances selected from the `PersonList` class.
PersonList.js	You'll extend `Ext.dataview.List` here to display the data in the data.json file and implement the `PersonStore` class.

Table 2.3 The list of files in the MyApp directory and the role of each in your application *(continued)*

File	Role
PersonModel.js	This class will define your data model for the person and provide the PersonList and PersonDetail widgets with the data for each record in the data.json file.
PersonStore.js	Here you'll extend `Ext.data.Store` and implement your `PersonModel` class. You'll configure this class to pull data from an online resource.

As a sibling to the MyApp directory you have app.html, which will launch your application.

That's pretty much all there is to it. Now that you have a clear picture of what you'll be constructing you can start getting your hands dirty.

2.5 Developing your app

Begin developing your application by defining the model (using the `PersonModel` class) that'll drive your data store extension, `PersonStore`. All of your classes will be inside the `MyApp` namespace.

You'll use the class `Loader` system, which means that you'll have to set up a dependency model, where one class will cause the automatic loading of another (figure 2.6).

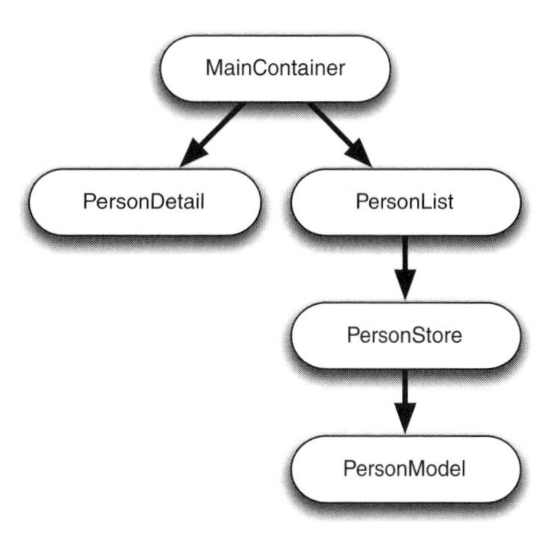

Figure 2.6 The class dependency model

> **Exposure to advanced concepts**
>
> You're going to use `Ext.define`, a method that allows you to declare classes. It's okay if you don't fully understand what's going on here. But if you want to learn about `Ext.define` take a peek at chapter 9. Be sure to come right back when you're done; we have a lot of work to do!

Let's begin by creating the `PersonModel` class. The code in the next listing will reside in MyApp/PersonModel.js.

Listing 2.3 The `PersonModel` extension

```
Ext.define('MyApp.PersonModel', {            ❶ Defines model
    extend : 'Ext.data.Model',               ❷ Extends Ext.data.Model
    config : {
        fields : [                           ❸ Configures data fields
            'city',
            'firstname',
```

```
                'lastname',
                'middle',
                'state',
                'street',
                'zip'
            ]
        }
});
```

To define your PersonModel class you employ Ext.define, specifying the full namespace for your MyApp.PersonModel ❶ extension. You instruct Ext JS to extend the Ext.data.Model class ❷ and configure the data fields ❸ for the model to use. These will be useful in mapping the inbound data from your remote web service and will be used to supply data to the PersonDetail extension.

That wraps up the PersonModel class file. We can now move on to adding content to the PersonStore.js file.

2.5.1 Creating the data store

Your next step is to define the data store extension, which will implement your PersonModel class. The following listing displays the code for PersonStore, which will reside in MyApp/Personstore.js.

Listing 2.4 The PersonStore extension

```
Ext.define('MyApp.PersonStore', {                          <-- ❶ Defines PersonStore
    extend   : 'Ext.data.Store',
    alias    : 'store.personstore',
    requires : [ 'MyApp.PersonModel' ],                    <-- ❷ Sets up class dependency
    config   : {
        autoLoad : true,
        model    : 'MyApp.PersonModel',                    <-- ❸ Implements PersonModel
        proxy    : {
            type    : 'jsonp',
            url     : 'http://extjsinaction.com/dataQuery.php',   <-- ❹ Uses external web data source
            limit   : 20,
            reader  : {
                type         : 'json',
                rootProperty : 'records'
            }
        }
    }
});
```

In listing 2.4 you define your PersonStore class ❶, an extension to Ext.data.Store. After instructing Ext JS to extend the data.Store class, you configure an alias parameter.

To have Sencha Touch automatically load the PersonModel class you'll need to document it as a dependency to the PersonStore class, which you do with the requires property ❷.

Next tell PersonStore to implement PersonModel ❸ via the model config and instruct it to use a JSON-P proxy that will request data from an online web service ❹. This data store will automatically load once instantiated.

You now have the data portion of your application code defined. You can now move on and configure your list extension.

2.5.2 Constructing the PersonList class

To create your `PersonList` class you'll have to extend `Ext.dataview.List`, which will require and implement the `PersonStore` class you defined previously. The contents of the `PersonList` class are shown in the next listing.

Listing 2.5 The `PersonList` class

```
Ext.define('MyApp.PersonList', {                    <-- ❶ Defines PersonList
    extend   : 'Ext.List',                               ❷ Configures
    xtype    : 'personlist',                         <--   XType alias
    requires : ['MyApp.PersonStore'],                <--
    config   : {                                         ❸ Requires
        allowDeselect : false,                             PersonStore
        itemTpl       : '{lastname}, {firstname}',
        store         : {                            <--
            type : 'personstore'                         ❹ Implements
        },                                                 PersonStore
        items         : [
            {
                xtype  : 'toolbar',
                title  : 'People',
                docked : 'top'
            }
        ]
    }
});
```

To define the `PersonList` class ❶ you extend `Ext.dataview.List` and set up an `XType` alias ❷ for lazy instantiation later on the inside of the `MainContainer` class you'll soon create. After the alias you set up the requirement for the `PersonStore` ❸ class and instruct Sencha Touch to load that class immediately after `PersonList` is loaded and defined.

The last block of code, the class config object, contains properties that allow you to configure how each instance of this class will behave. For example, you set the `itemTpl` property as a string that contains tokens that'll render first and last names, as defined in your data store's model. This works because you configured your `PersonList` class to use your `PersonStore` class via lazy instantiation ❹.

Why so lazy?

Lazy instantiation is a pattern in which you define a plain JavaScript object and use it to configure an instance of some class. We'll talk about lazy instantiation in chapter 3 when we discuss `XTypes`.

The `items` configuration property allows you to configure a toolbar that will be docked to the top of your `List` extension, providing a nice area for a title. Figure 2.7 shows what it'd look like if you rendered it onscreen by itself.

Your `PersonList` class is complete. Next up is the PersonDetail screen.

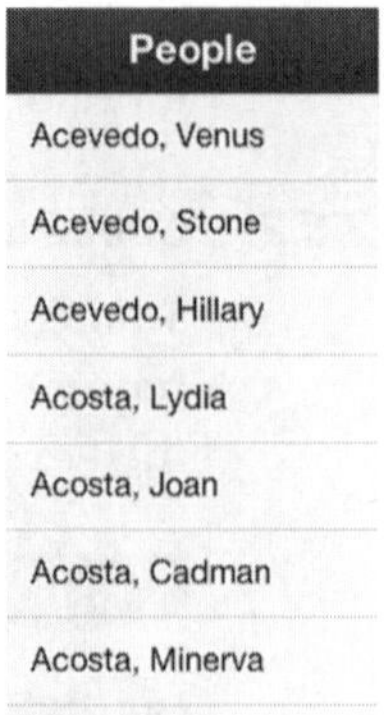

People
Acevedo, Venus
Acevedo, Stone
Acevedo, Hillary
Acosta, Lydia
Acosta, Joan
Acosta, Cadman
Acosta, Minerva

Figure 2.7 The `PersonList` class rendered by itself

2.5.3 *Building PersonDetail*

The extension to the Sencha Touch form `Panel` class, which will be known as `PersonDetail` in your code base, will be responsible for constructing all the fields required to display the details of a record selected in the `PersonList` class.

The following listing contains the code for this new class. This will be the longest listing in this chapter, simply because the input fields require a lot of code, relatively speaking.

Listing 2.6 The `PersonDetail` class

```
Ext.define('MyApp.PersonDetail', {                  <-- 1 Defines PersonDetail
    extend : 'Ext.form.Panel',
    xtype  : 'persondetail',                          <-- 2 Configures XType
    config : {
        items: [
            {
                xtype        : 'fieldset',                <-- 3 Configures input fields
                defaultType  : 'textfield',
                defaults     : { labelWidth : 100 },
                items        : [
                    {
                        label : 'First',
                        name  : 'firstname'
                    },
                    {
                        label : 'Last',
                        name  : 'lastname'
                    },
                    {
                        label : 'Street',
                        name  : 'street'
                    },
                    {
                        label : 'City',
                        name  : 'city'
                    },
                    {
                        label : 'State',
                        name  : 'state'
                    },
                    {
                        label : 'Zip',
```

```
                        name  : 'zip'
                    }
                ]
            },
            {
                xtype  : 'toolbar',
                title  : 'Person Details',
                docked : 'top'
            }
        ]
    }
});
```

In listing 2.6 you define the `PersonDetail` class ❶ and its `XType` alias ❷. The longest section of this listing defines the child items ❸ for this extension. Here you define all of the input fields to display the data and a toolbar (docked at the top of the widget) that contains a simple title. Figure 2.8 illustrates what this widget looks like rendered onscreen alone with data in it.

Person Details	
First	Lydia
Last	Acosta
Street	2446 Ipsum Street
City	Huntsville
State	GA
Zip	97985

Figure 2.8 The PersonDetail widget rendered onscreen

Your `PersonDetail` class is now complete and ready to be stitched in. Next you'll create the `MainContainer` class, which will allow you to tie everything together.

2.5.4 *Setting up the MainContainer class*

As we mentioned a bit earlier, the `MainContainer` class is going to be the master UI class for your mini-application. This widget will be responsible for rendering the PersonList and PersonDetail widgets onscreen and will tie in the simple left-to-right workflow using event listeners.

The next listing contains the code for this `MainContainer` class and has some of the most advanced Sencha Touch implementation code you've seen so far.

Listing 2.7 The `MainContainer` class layout

```
Ext.define('MyApp.MainContainer', {          <-- ❶ Defines MainContainer
    extend   : 'Ext.Container',
    requires : [                             <-- ❷ Requires PersonList,
        'MyApp.PersonList',                        PersonDetail
        'MyApp.PersonDetail'
```

```
    ],
    config : {
        layout : {
            type   : 'hbox',
            align  : 'stretch'
        },
        items : [                                          ◁─┐ Implements
            {                                                │ PersonList,
                xtype  : 'personlist',                       3 PersonDetail
                itemId : 'list',
                width  : 200,
                style  : 'border-right: 1px solid #999'
            },
            {
                xtype  : 'persondetail',
                itemId : 'detail',
                flex   : 1
            }
        ],
        listeners : {
            select : {                                     4 Listens for
                fn       : 'onListSelect',              ◁─┘ select event
                delegate : '> #list'
            }
        }
    },
    onListSelect : function(list, record) {              ◁─┐ Displays
        this.getComponent('detail').setRecord(record);    5 data
    }
});
```

You first define your `MainContainer` ❶ class in listing 2.7. Because you'll be instantiating this class directly there's no need to configure an `XType` alias for it. To have Sencha Touch automatically load your `PersonList` and `PersonDetail` classes you must list them in the `requires` configuration block ❷.

Next you implement your PersonList and PersonDetail widgets by means of lazy instantiation with their `XTypes` ❸. You configure each widget with a special `itemId`. Doing so will allow you to do things such as configure event listeners and gain references to these components later on.

To tie into the selection from `PersonList` and load the selected record into the `PersonDetail` class you configure a `listeners` object, which defines the event to listen to (`select`), what function to call (`onListSelect`) ❹, and which child to hook the event into (`PersonList`). This uses the new power of the Sencha Touch 2 event system, where you can define event handlers via `String` representation of method names and use `ComponentQuery` selector syntax to guide event handler registration.

The `onListSelect` method ❺ will always be called within the scope of your instance of `MainContainer`. Therefore, you can use the `this.getComponent`, a `Container` method, to gain a reference to the `PersonDetail` instance via `itemId` and execute its `setRecord` method, passing in the model instance (`record`) as the only argument. All of this will cause the `PersonDetail` class to bind to the instance of the record and display its data in the input fields.

All of your classes are defined and ready to roll. Now you can render your app.

2.5.5 Rendering your application

To render your application you'll need to add HTML and JavaScript to your app.html file. The following listing contains the contents of that file.

Listing 2.8 The contents of app.html

```
<!DOCTYPE html>
<html>
<head>
    <meta http-equiv="Content-Type"
         content="text/html; charset=utf-8">
    <title>Sencha Touch in Action CH2 app</title>
    <link rel="stylesheet"
         href="../../sencha-touch-2.0.0/resources/css/sencha-touch.css"
         type="text/css">
    <script type="text/javascript"
         src="../../sencha-touch-2.0.0/sencha-touch-all-debug.js">
    </script>

    <script type='text/javascript'>
        Ext.Loader.setConfig({                              ❶ Enables class loader
            enabled : true,
            paths   : {
                MyApp : 'MyApp'                             ❷ Configures namespace and path
            }
        });

        Ext.require([                                       ❸ Dynamically loads MainContainer
            'MyApp.MainContainer'
        ]);

        Ext.application({
            launch : function () {
                Ext.create('MyApp.MainContainer', {         ❹ Renders MainContainer
                    fullscreen : true
                })
            }
        });
    </script>

</head>
<body>
</body>
</html>
```

Listing 2.8 contains everything you need to render the entire application. Aside from the HTML that defines the page and loads Sencha Touch and the necessary CSS, you have a script tag that does a bit of work to bootstrap things and render the application.

The first step to render your app is to enable the Sencha Touch class `Loader` via the `Ext.Loader.setConfig` method call ❶. You also have to tell `Loader` in which directory to look for the class files in the `MyApp` namespace ❷, and you do this via the `paths` property.

To render the application you call `Ext.require` and pass your `MainContainer` class ❸. Doing so allows Sencha Touch to load this class using script tags, a process that's considered best practice for performance, and will delay the "ready state" for the browser. Because you have your class requirements well defined, `MainContainer` is the only class you'll need to instruct Ext JS to load.

The last step to render your application onscreen is to call `Ext.application` and pass an `onReady` ❹ method to it. Setting up this method will allow your application to launch at the correct time in the browser's page load cycle. You need this because your app will attempt to manipulate the DOM before it's fully built by the browser and, without this, you'd get exceptions and never leave the launch pad.

Inside the `onReady` function you call `Ext.create` and pass a `String` representation of your classname. With Sencha Touch `Ext.create` replaces the use of the JavaScript `new` keyword because it plays nicely with the `Loader` system.

> **New is so old!**
> We realize that this may seem foreign to you at this stage in the game. It did to us at first as well! If you peek at chapter 9, where we dive into the class `Loader` system, you'll gain an appreciation for `Ext.create`.

If you load your application in a WebKit browser (we typically choose Google Chrome) you can see the class system at work. Figure 2.9 shows you what we saw when we peeked at the WebKit debug tools.

Looking at the WebKit Inspector Network tab, you can see that immediately after Sencha Touch was loaded the `MainContainer` class was loaded. Next, `PersonList` and then `PersonDetail` were loaded. The `Loader` system inspected the requirements for `PersonList` and loaded `PersonStore`, and then loaded its requirement, `PersonModel`.

All of this sets the stage for your application to function. Figure 2.10 shows what it looks like rendered in Chrome after we've selected the first record in the `PeopleList` UI.

After rendering your application you can see that selecting a record from the PeopleList widget on the left will result in the details of that record being displayed on the right.

This read-only application sets the stage for having full-blown create, read, update, and delete (CRUD) against the data. You could create buttons for Create, Update, and Delete operations, but we think this is a good time to stop. We hope this example has whetted your appetite to learn more of what's happening under the hood in this framework.

In the next chapter we'll cover understanding critical concepts, such as lazy instantiation.

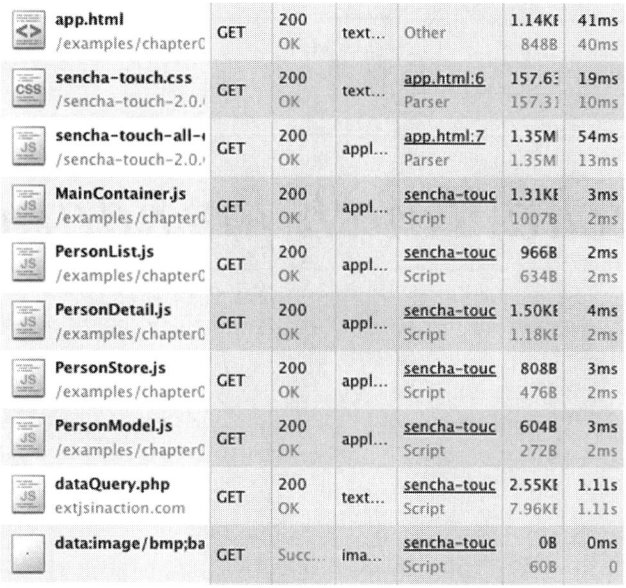

Name / Path	Method	Status	Type	Initiator	Size	Time
app.html /examples/chapterC	GET	200 OK	text...	Other	1.14KE 848B	41ms 40ms
sencha-touch.css /sencha-touch-2.0.	GET	200 OK	text...	app.html:6 Parser	157.6: 157.3	19ms 10ms
sencha-touch-all- /sencha-touch-2.0.	GET	200 OK	appl...	app.html:7 Parser	1.35M 1.35M	54ms 13ms
MainContainer.js /examples/chapterC	GET	200 OK	appl...	sencha-touc Script	1.31KE 1007B	3ms 2ms
PersonList.js /examples/chapterC	GET	200 OK	appl...	sencha-touc Script	966B 634B	2ms 2ms
PersonDetail.js /examples/chapterC	GET	200 OK	appl...	sencha-touc Script	1.50KE 1.18KE	4ms 2ms
PersonStore.js /examples/chapterC	GET	200 OK	appl...	sencha-touc Script	808B 476B	3ms 2ms
PersonModel.js /examples/chapterC	GET	200 OK	appl...	sencha-touc Script	604B 272B	3ms 2ms
dataQuery.php extjsinaction.com	GET	200 OK	text...	sencha-touc Script	2.55KE 7.96KE	1.11s 1.11s
data:image/bmp;ba	GET	Succ...	ima...	sencha-touc Script	0B 60B	0ms 0

Figure 2.9 We can see the Sencha Touch Loader working as it loads the classes.

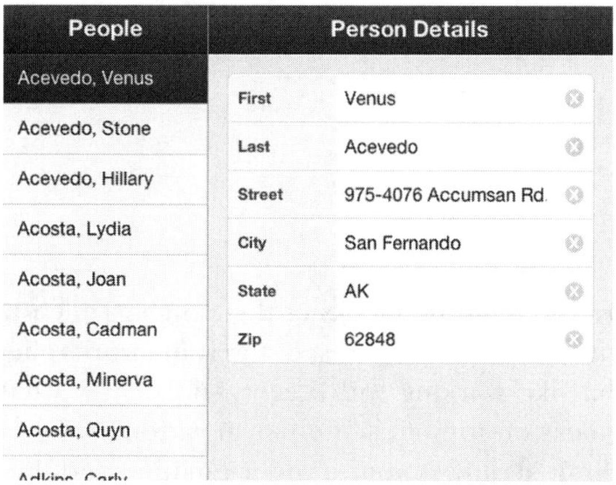

Figure 2.10 The application rendered in Google Chrome

2.6 Summary

In this chapter you downloaded the framework and delved into its contents. Then you created a simple "Hello World" example, implementing the MessageBox widget.

Next you created a simple application. In this exercise you were exposed to critical bits of knowledge like defining classes, `XTypes`, class dependencies, and event delegation. You rendered your application in Google Chrome and learned how your classes were dynamically loaded, and you saw how to run your application.

In the next chapter you'll begin your journey into the heart of Sencha Touch, beginning with a thorough look into the component life cycle and a discussion of the various layouts in the framework. This will help you set the foundation for further exploration.

3 Sencha Touch foundations

This chapter covers

- Getting to know the Component model
- Exploring the component life cycle
- Learning utility methods to create components

Sencha Touch *components* are the building blocks of the Sencha Touch universe. They make up the foundation upon which the rest of the framework relies. Working with components is a bit like working with Legos—no matter what you're building, you always stack blocks on top of each other in various ways. The wondrous things you create are limited only by your own imagination and the rules of the framework.

To grasp exactly how Sencha Touch enforces these rules and how components fit together you have to understand the underlying Component model and its relationship to components. In this chapter you'll take a close look at the component life cycle and learn why you should care about it. In addition you'll see the various ways of creating components and look at the pros and cons of each approach. But first you need to build a good foundation to expand from, so let's take a look at the Component model.

3.1 One Component model to rule them all

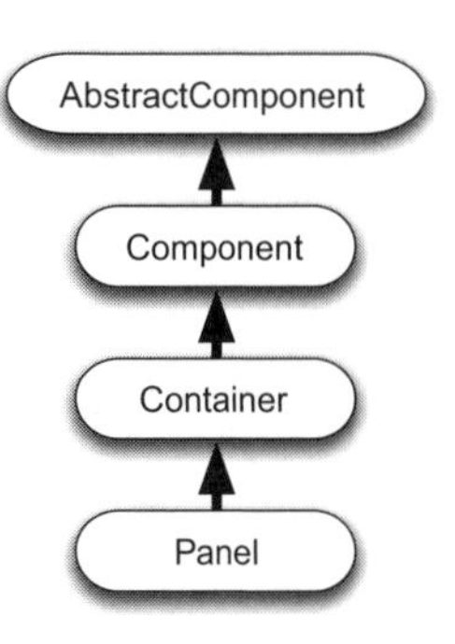

Figure 3.1 The common components in the Sencha Touch Component model, depicted with their hierarchies and subclassing

Part of the magic behind Sencha Touch, and what makes it so special compared to other frameworks, is the way in which UI widgets are structured and fit together almost seamlessly. In every UI widget, if you go far enough up the inheritance chain, you always arrive at the same common ancestor, the `Ext.AbstractComponent` class. Every UI piece in Sencha Touch is a subclass of `Ext.AbstractComponent`, something that's made possible via the Sencha Touch Component model, a centralized model that governs all component-related tasks. `Ext.AbstractComponent` is the base of the Component model but shouldn't be used directly. Instead you should use `Ext.Component`. The Component model provides a set of rules and behaviors that all children of `Ext.AbstractComponent` have to follow in some way or another. Figure 3.1 depicts the common components in the Component model, showing just how many UI pieces subclass Component, either directly or indirectly. Fret not if figure 3.1 seems overwhelming at first glance. It simply provides a big picture overview of all of the Sencha Touch components and can serve as a good reference point to return to as you read later chapters or venture into your own development endeavors.

Figure 3.1 shows the more common components in Sencha Touch 2 that you'll most likely use in everyday applications. These aren't all the components that Sencha Touch 2 has. Next in figure 3.2 you'll see the entire hierarchy of the Component model in Sencha Touch 2, which may look overwhelming but is a great depiction of where each component is in the hierarchy.

One thing that might not be immediately apparent when looking at figures 3.1 and 3.2 is what exactly the benefits of the Component model are. First, knowing what components another component extends is valuable knowledge. The rules the Component model has to adhere to ensure dependability. If you know what behavior a Container has, you can depend on the knowledge that its subclasses, `Panel` for instance, will most likely behave in the same way (for example, events will fire in a certain order). This dependability is one of the most overlooked benefits of the Component model; whether or not you know it, you'll be relying on this every day you're developing with Sencha Touch.

Second, knowing exactly how each UI widget is going to behave introduces predictability and stability into the framework. Whether you're dealing with native Sencha Touch widgets or components created by third-party developers, you can always count on specific events and behaviors to occur at predetermined points.

What makes all of this possible is the thing that governs how a component is instantiated, rendered, and destroyed: the component life cycle.

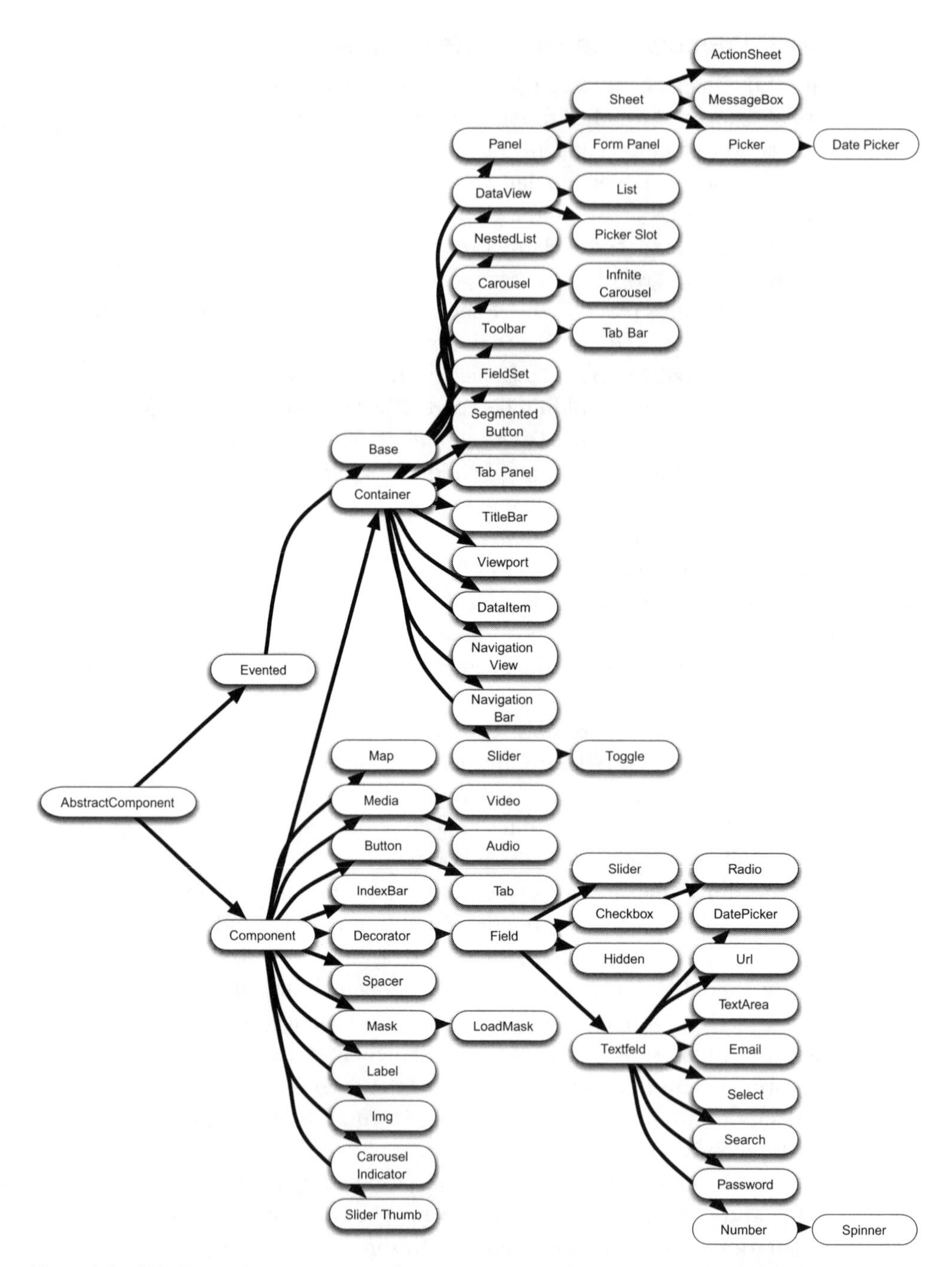

Figure 3.2 This illustration of the Sencha Touch class hierarchy maps out all the subclasses of `Ext.AbstractComponent` and depicts how widely used the Component model is in the framework.

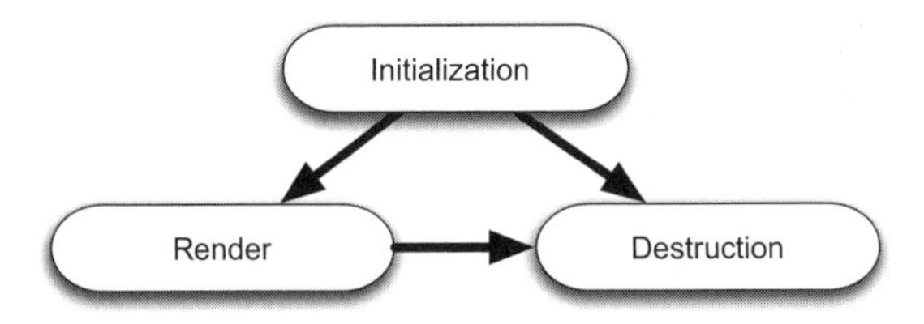

Figure 3.3 The Sencha Touch component life cycle always starts with the initialization phase and ends with the destruction phase. The component doesn't need the render phase to be destroyed.

3.2 *Introducing the component life cycle*

Each component in the Sencha Touch universe shares a common ancestor, `Ext.Component`. Just as in genetics, this ancestor passes down some of its traits. One of those traits is the component life cycle, a set of rules and behaviors that determines what happens when a component is "born," what happens while it "lives," and what happens when it "dies."

Given that the average app created by the average user frequently consists of out-of-the-box components without too much customization, the need to study all the gory details behind the component life cycle can be assessed by each developer. For those "average" developers, the component life cycle will be one of those things that they hear about on the forums and that they'll most likely have a high-level understanding of—which tells them that a component is created during the initialization phase, rendered during the render phase, and destroyed during the destruction phase. Knowing a lot of details beyond that isn't a requirement, at least not for the average app. For those developers who wish to venture into creating their own custom components and plug-ins that extend the core functionality of the framework this somewhat dry topic will provide a goldmine of information. Quite a few steps take place at each of the three phases in the life cycle (see figure 3.3).

Everything must have a beginning, so let's start with the beginning of each component: the initialization phase.

3.2.1 *Initialization/instantiation phase*

The initialization phase is when a component is born. Any necessary prerender actions like configuration settings and HTML structure creation take place during this phase. The exact steps break down as shown in figure 3.4.

Let's explore each step of the initialization phase. The numbers in figure 3.4 correspond to each step in the following list:

1. *The component configuration is applied*—When creating an instance of a component you pass it a configuration object that contains all the necessary knobs and dials to tell the component what to do and how to behave.
2. *A unique ID is created*—Each component within your app must have a unique identifier. This requirement is dictated by the `Ext.ComponentManager`, because all components are automatically registered with the `ComponentManager`.

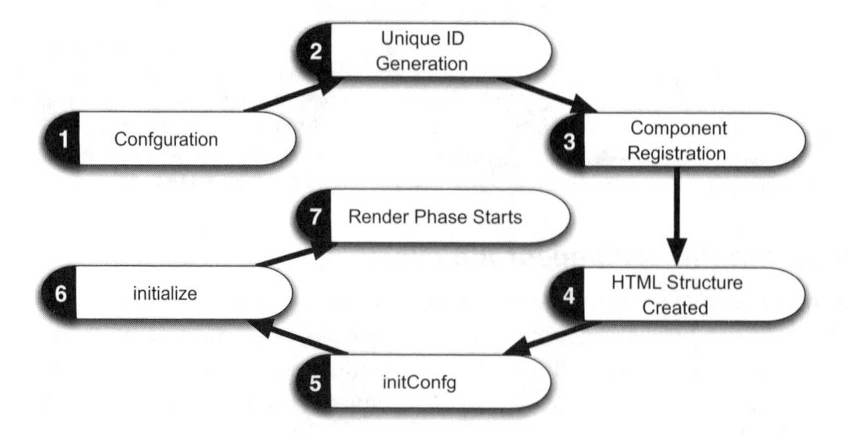

Figure 3.4 The initialization phase of the component life cycle executes important steps such as component registration and calling the `initConfig` and `initialize` methods. It's important to remember that a component can be instantiated without being rendered.

There are two ways a component can come by its ID:

- The first is for Sencha Touch to automatically generate an ID, which is the default and preferred behavior when no ID is supplied during component instantiation. Sencha Touch–generated IDs usually take the form of `ext-component-1`, `ext-component-2`, and so forth. The format is `ext-{xtype}-{number}` where `{xtype}` is the XType of the component and `{number}` is an incremental number.
- The second is for you (the developer) to supply an ID during instantiation, thus providing more control over what the ID should be. The ID can be any string you desire; the only caveat to keep in mind is that the `ComponentManager` doesn't allow duplicate IDs within the entire application. This means it's up to you to keep track of custom IDs you set and ensure there are no duplicates. Whenever a component with an already existing ID is registered the `ComponentManager` overwrites the already registered version with the newly supplied one. Nuking existing components causes undesirable effects and is a pain to debug, so be careful about this.

ItemId to the rescue

One easy way to alleviate having to track countless IDs across your entire application is to use the `itemId` property for IDs assigned by you and to let Sencha Touch automatically assign the `id` whenever possible. The `itemId` has to be unique within a container, meaning you could have two `Panel` instances, each containing a `Textfield` with an `itemId` of "myTextField," and not run into any collision issues.

> **ComponentQuery**
> A more flexible and robust way is to make use of `Ext.ComponentQuery`, taking advantage of the hierarchy of your application. You can go up and down the hierarchy searching `XTypes` and properties. The caveat is there's a small performance hit when using this, but today's devices are getting more powerful every day.

3. *The component is registered with the* `ComponentManager`—Each instance of a component is automatically registered with the `Ext.ComponentManager` using its unique ID. If you know this unique ID you can utilize the `Ext.ComponentManager` to perform a lookup of the component and get a handle on it for interaction via the `Ext.getCmp` function.
4. *The HTML structure is created*—Each component has its own HTML structure so that it can display properly. This step creates the HTML structure, but it's not placed in the DOM to be rendered yet. Sencha Touch 2 creates this structure so it can do a one-batch DOM write instead of doing multiple DOM writes. This approach helps with performance—one of the things that'll slow your app the most is DOM write—so minimizing the number of writes is a good way to improve performance. Remember, you're still in the initialization phase; the component won't be rendered until the render phase later in the life cycle.
5. `initConfig` *is executed*—One of the new key features in Sencha Touch 2 is the `config` object. The `config` object automatically creates a `getter` and `setter` method for each `config` object item. `initConfig` is the method that every component (and every class) has, and it performs the creation of the `getter` and `setter` methods. A `getter` method will return the property value that was in the `config` object. Not only will the `setter` method set the property value, but it will also use `apply` and `update` methods. When you execute the created `setter` method it first looks for an `apply` method and executes it, passing in the new and old values as arguments. The `apply` method is great for transforming the value or providing validation, and it must return a value. After the `apply` method is executed the property value is then set. Then the `update` method is fired, passing the new and old values. The `update` method takes action on setting a new value, like updating some other component or DOM element.
6. `initialize` *is executed*—The `initialize` method is used to add additional configuration before the render phase starts. `initialize` is the last template method that you can override for subclassing before the component is rendered. This can provide a last chance to change items or apply event listeners. Note that some Sencha Touch components use `initialize`, so if you need to override this method you'll usually have to execute `callParent` to call the superclass's method.

7. *The component is rendered*—If the `fullscreen` config is set to `true` the component will fire the `fullscreen` event, which the global `Ext.Viewport` component listens to, and the Viewport will automatically add the component to itself, triggering the rendering.

This phase of a component's life is usually the swiftest, due to the fact that all the work is done behind the scenes in JavaScript, and no rendering is necessary up to this point. It's particularly important to remember that a component doesn't have to be rendered in order for it to be destroyed.

3.2.2 Render phase

The render phase provides visual feedback that a component has been successfully instantiated. If the component encountered any issues during the initialization phase it's unlikely to render correctly or at all. Unless the `fullscreen` property is specified rendering of the component is deferred until the component is added as an item of a container. We'll talk more about containers in the next chapter.

If your component is a child of a container the container's layout usually takes care of doing the rendering. The following code shows how to use the `fullscreen` property to have `Ext.Viewport` automatically add the component as an item:

```
var myComponent = new Ext.Component({
    fullscreen : true,
    html       : 'testing'
});
```

In this example you instantiate an instance of `Ext.Component` and create a reference to it, `myComponent`. After the instance is instantiated it sees the `fullscreen` property is set to `true` and will then fire the `fullscreen` event on itself. `Ext.Viewport` listens for the `fullscreen` event and when a component fires this it'll automatically add that component as an item of itself.

In Sencha Touch 1 you usually used the `renderTo` or `fullscreen` config, which would render that component to a particular element, or you'd add a component to a container and that would then render the component within the container. In Sencha Touch 2 you can still render components to an element, but you're going to add components as items of containers 99% of the time.

Also in Sencha Touch 1 the layouts were JavaScript-based; they didn't perform well and weren't 100% accurate. In Sencha Touch 2 layouts are CSS-based (see figure 3.5), so they're going to be highly accurate and perform very well. In the case of an orientation change the layouts are usually finished before the browser has finished updating its size, so you can't get much faster than that.

The component rendering is split into multiple steps, as you see in figure 3.5. The number for each step corresponds to the number in the list that follows. There are a few steps that are big but important. Let's explore the steps of the render phase:

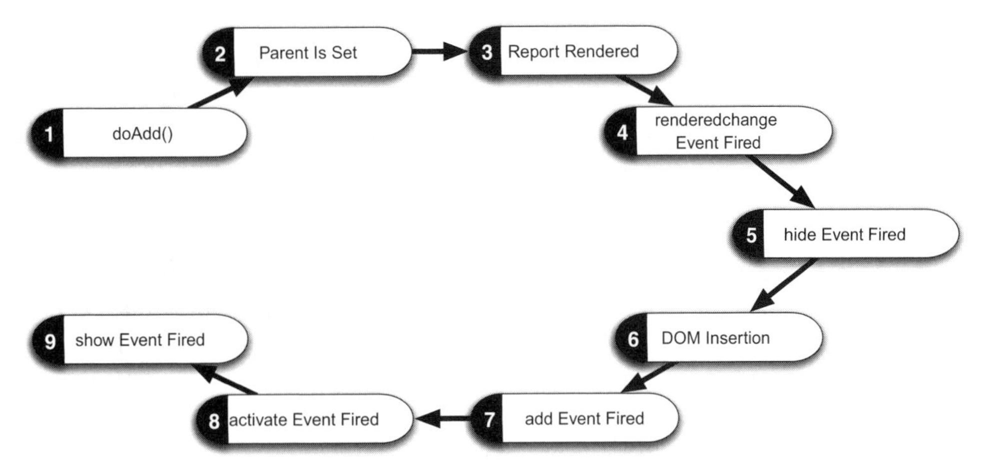

Figure 3.5 The render phase of a component is simple when layouts are CSS-based.

1. `doAdd` *is called*—The `doAdd` method is called, adding the component to the container as an item. This method doesn't render anything by itself, but it kicks off the rendering process.
2. `setParent` *is called*—The component needs to have a defined parent. This method checks to see if the `parent` property on the component is set, and if it is the method automatically removes the component from the current parent. Then it'll set the `parent` property to the container that the component is being added to.
3. *The component is reported rendered*—At this point the component reports that it's rendered. This step is a little tricky, because the component isn't rendered in the sense that it's in the DOM. The HTML structure was already created back in the initialization phase, so it's rendered outside of the DOM. It's a little crazy to think that way, but it gives the framework a chance to change the HTML before it's inserted into the DOM. Remember from earlier that DOM writing is slow. Doing it this way saves a few crucial milliseconds.
4. *The* `renderedchange` *event is fired*—Because the component now reports that it's rendered, the `renderedchange` event is fired. If you use this event to manipulate the element of the component you need to be aware that the component isn't in the DOM at this point. This is important: you can do things to the component, but your options may be limited because it's not in the DOM.
5. *The* `hide` *event is fired*—This is another important step. The component is still not in the DOM, but it's getting hidden. But why? When the component is finally inserted into the DOM it's hidden in order for the layouts to correctly catch up so you don't see the browser jerking things around. This approach results in a much better user experience. It's like the saying, "Sometimes you have to take a step back to go forward." You'll see in the following steps that there are still things that are going to happen to the component.

6 *The component is inserted in the DOM*—It's been a long road, but the component is now finally inserted into the DOM! The component is still hidden, but it's there. You have a few more things to do and then it'll be made visible.
7 *The* `add` *event is fired*—The parent of the component will now fire an `add` event. This isn't a step for the component because its parent is the one that fires the `add` event, but it's part of the rendering process of the component and you should know about it as you develop your application.
8 *The* `activate` *event is fired*—This event will fire only if the container of the component doesn't already have an active item. If the component is to be made the active item it'll have the `activate` event fired. This event will also make the component visually active so it's no longer hidden, but do remember that the event is fired before the component is visually shown. The `show` event is fired after the component is visually shown at this point.

The render phase consists of a lot of events. Because the process is event-driven the code remains flexible. You could handle this phase by executing methods, but the same events the framework uses to render a component can also be listened to when you're developing an application; methods are much harder to hook into.

You've seen how a component is created in the initialization phase and optionally rendered in the render phase. The last part of the component life cycle is to destroy the component in the destruction phase.

3.2.3 Destruction phase

The death of a component is a crucial step in its life. Destroying the component is performed via the `destroy` method, which takes care of critical tasks such as removing the component and any children from the DOM tree, purging event listeners, and unregistering the component from the `Ext.ComponentManager`. The component's `destroy` method can be called by a parent container or manually by your code. Figure 3.6 illustrates the steps that go into the destruction phase.

With our Lego blocks from childhood it was always harder to build something than to tear it down. The destruction phase is no different in that regard. Here are the steps that make up the final stage of a component's life. Keep in mind that each step corresponds to a step in figure 3.6.

1 *The component is removed from the parent*—If the component is an item of a container, it'll be removed from the container.
2 *The* `renderedchange` *event is fired*—The `renderedchange` event is fired, telling the component that it'll no longer be rendered. Note that the event is fired before the component is removed.
3 *The DOM is removed*—Unlike the render phase, where you had to wait for the DOM to reflect the component, the destruction of the component removes the DOM elements early on.
4 *The* `remove` *event is fired*—Now is the time for the container to announce that the component has been removed by firing the `remove` event. The `remove` event

is an important step in the destruction phase, even though the event isn't fired on the component itself.

5. *References are destroyed*—Each component has a `referenceList` property that's an array of strings, with each string being a property on the component. For example, all components will have the string `'element'` in the `referenceList` array, and every component has an `element` property that's an `Ext.dom.Element` of the component's DOM. This step loops through each item and executes the `destroy` method on each item. Even though the DOM elements for the component have been removed there are still `Ext.dom.Element` references to them and, if not destroyed, they'll cause a memory leak.

   An advanced technique when you're creating custom components is to utilize this `referenceList` array to clean up references. You just have to make sure it has a `destroy` method or you'll get a JavaScript error.
6. *The component is deregistered*—The reference for the component in the `Ext.ComponentManager` is removed.
7. *Listeners are purged*—All listeners on the component are cleared and removed from the component. If this step didn't happen the leftover listeners would be considered a memory leak and would quickly add up and slow down your application.

Now that you've endured the long and arduous road through the component life cycle, we want to warn you not to underestimate the importance of the three different phases, especially when you're venturing into creating your own components. Many developers have dismissed the destruction phase, only to leave a continuously polling data store hanging around to cause havoc.

The last part of this chapter shows you how to create and manage components.

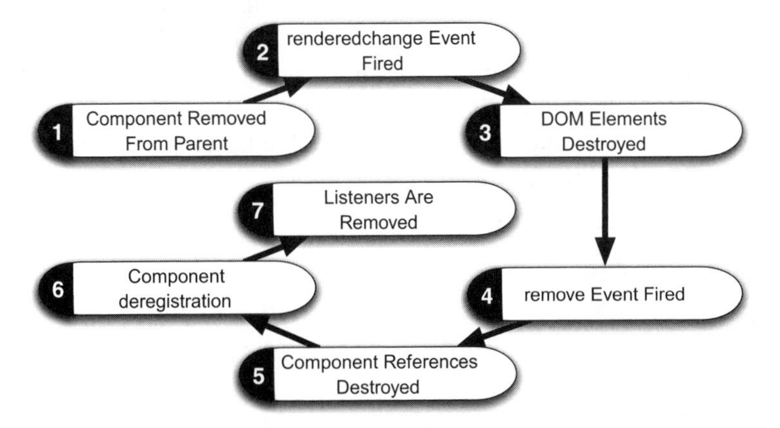

Figure 3.6 The destruction phase of a component is equally as important as the initialization phase. This is where cleanup of event listeners and DOM elements, as well as deregistration and removing of elements, happens—all in the spirit of keeping your application running like the well-oiled machine that it is.

3.3 XTypes and the ComponentManager

Each component within the Sencha Touch framework can be created in one of three ways: direct instantiation, direct-loaded instantiation, and lazy instantiation. So what's the difference? Direct instantiation will instantiate a component right away when the browser executes the code. Direct-loaded instantiation is similar to direct instantiation in that the component will be created right away, but only after checking if the class is loaded. Sencha Touch can dynamically load components, so in order to create a component the file has to be loaded and then the component can be instantiated. Lazy instantiation uses a configuration object. Once the component is required to be instantiated the component will then be instantiated using that configuration object.

3.3.1 Examples of instantiations

Direct instantiation will look familiar to most programmers; most languages use it predominantly. In this case the component is created explicitly via its constructor, providing instant access to the component and all of its properties, including the component's DOM structure once it renders.

In practice it looks something like this:

```
var myPanel = new Ext.Panel({
    modal             : true,
    hideOnMaskTap     : true,
    centered          : true,
    scrollable        : 'vertical',
    width             : 300,
    height            : 200,
    styleHtmlContent  : true,
    html              : '<h2>Hello Readers</h2>'
});
```

The direct-loaded instantiation method makes use of `Ext.create`. `Ext.create` can create a Sencha Touch component. This is similar to the direct instantiation method where it creates the component immediately, but here it uses `Ext.Loader`. If the component isn't loaded, `Ext.Loader` will load that class and then create the component like direct instantiation does via the constructor. We'll take a look at `Ext.Loader` a bit later. Here's an example of using `Ext.create`:

```
var myPanel = Ext.create('Ext.Panel', {
    modal             : true,
    hideOnMaskTap     : true,
    centered          : true,
    scrollable        : 'vertical',
    width             : 300,
    height            : 200,
    styleHtmlContent  : true,
    html              : '<h2>Hello Readers</h2>'
});
```

Sencha Touch also allows for the use of lazy instantiation, an alternative method to direct and direct-loaded instantiation that uses an `XType` to represent the component:

```
var myPanel = {
    xtype             : 'panel',
    modal             : true,
    hideOnMaskTap     : true,
    centered          : true,
    scrollable        : 'vertical',
    width             : 300,
    height            : 200,
    styleHtmlContent  : true,
    html              : '<h2>Hello Readers</h2>'

};

Ext.ComponentManager.create(myPanel).show();
```

Instead of explicitly creating a new instance of a component, a plain JavaScript object containing a special string property named an `XType`, along with the other various configuration values for the class, is used in its place. The `XType` value serves as a unique identifier linking the string value to a Sencha Touch component registered with the `Ext.ComponentManager`, a singleton class that tracks all components and their `XType`s. This link tells the `ComponentManager` which component to instantiate when an `XType` is encountered. The reason the previous code sample works is because every single UI widget is registered with the `Ext.ComponentManager`, and the code uses `'panel'` as the `XType`, which corresponds to the `Ext.Panel` component.

Whenever a component is initialized the `ComponentManager` automatically checks whether the component contains any child items, and for each child item checks whether the `XType` property is set. If it is, a new instance of the class that corresponds to the `XType` is created where the child component should be.

3.3.2 The pros and cons

Now that you know the ways to instantiate a component, when should you use one over the other? First let's look at some pros and cons of using each method in order to make a decision about what to use.

Direct instantiation will create the component right away, which is good, right? Yes and no. It's good because that component and all the properties and methods are available right away, but the problem is that in some situations you don't need to have that component instantiated right away. Direct-loaded instantiation has the same issue but could create an Ajax request to load the component's source file, which could delay start-up time and thus give the user the illusion that the application is running slowly, and performance would take a negative hit.

If that component isn't being used then it may not need to be instantiated and take up precious memory on the mobile device; this is where lazy instantiation may be a better choice. When you don't need to have all those properties and methods available you can use lazy instantiation to have only a small object stored in memory. Once the component is needed it'll be instantiated into a full component, thus saving memory on that mobile device, and your application will perform better. Later in chapter 9 you'll learn about creating your own class using `Ext.define` and in the `config` property you'll use lazy instantiation.

Another benefit of using lazy instantiation is that your code will be cleaner and more elegant. Child items can be declared in line versus creating them one by one, resulting in more streamlined code. This configuration object can be assembled dynamically by your code and even be retrieved from a backend by Ajax.

So lazy instantiation sounds fantastic, doesn't it? Sometimes, when something sounds too good to be true, it can be. Using lazy instantiation defers both the instantiation phase and the render phase of the component life cycle, resulting in a loss of immediate access to the component, something that isn't a problem when using direct or direct-loaded instantiation.

If you need to take action on a component, it must be instantiated. If the component hasn't been transformed into a full component you can't do anything with it. You always need to be aware of this limitation in your application. Is that component going to be a component or is it going to be a configuration object?

There are many things to take into consideration, but don't be overwhelmed. The choice between direct instantiation and direct-loaded instantiation is simple: could that component's source file not be loaded? If there's a chance it might not be loaded, then direct-loaded instantiation is the way to go; if it's going to be loaded, then using `Ext.create` may cause a little overhead that isn't needed and you can just go with direct instantiation. The decision between direct/direct-loaded instantiation and lazy instantiation comes down to whether or not the component is needed when the line of code is executed by the browser's JavaScript engine. Lazy instantiation can allow you to boost performance, but it can also create more work because it's not a component. If you can get away with using lazy instantiation we recommend it over the other two options.

3.4 Summary

This chapter may not be the most glamorous chapter, but it's important. Before you can build the walls of a house you must have a stable foundation. You saw the entire Component model, checked out the component life cycle and its three phases, and explored how to create a component. You also learned what to think about when deciding how to instantiate a component in your application. So far you've only scratched the surface of what Sencha Touch provides and what's possible. If you aren't yet comfortable with the material covered in this chapter it might be prudent to move on and periodically circle back to this chapter as a refresher. Ready for some more excitement? Next we're going to take a look at containers and some different panels that liven up Sencha Touch components.

Part 2

Building mobile user interfaces

By now you've obtained core knowledge that's essential to implementing the Sencha Touch framework. In this part you'll start to see the various widgets that comprise Sencha Touch. You'll also have an opportunity to look at how data-driven views work.

We kick things off with chapter 4, where you get an opportunity to look at how containers and layouts work to arrange your UI. Chapter 5 is focused on how components are "docked" and how to use toolbars effectively. Chapter 6 shows you how to use sheets and other widgets to focus the user's attention. In chapter 7 you learn how to use data stores to drive views. Chapter 8 focuses on getting data from the users with form panels and various input fields. We close out this part with chapter 9, where you learn how to use Google Maps, audio, and video in your applications.

By the end of this part you'll have a solid foundation of how Sencha Touch widgets work and how to use each of them effectively.

4 Mastering the building blocks

This chapter covers

- Exploring the Sencha Touch Container model
- Learning utility methods to arrange and manage widgets
- Learning to float and drag panels
- Implementing the TabPanel

In the previous chapter we covered a lot of foundational topics, including the Component model and component life cycle, and most important, how to create components. Armed with all this knowledge, you can now see how Sencha Touch uses these new concepts in some of the major UI components and thus master such topics as containers, panels, and TabPanels—exactly the things covered in this chapter. We'll explore how components are managed within containers and how to arrange components in various ways using the provided layouts. In the process we'll look into building a simple wizard-style interface, and explore how to build a TabPanel that looks similar to the typical iOS apps with icons at the bottom.

First things first: let's begin by diving into the anatomy of the `Container` class.

4.1 Containers: Mounting our UI workhorse

The container is the workhorse widget of Sencha Touch because it's typically deployed to display and help organize your application layout. It provides the foundation for components to manage their unruly children and facilitates ways to add, insert, remove, and query child components. The container is an extremely versatile widget and is even deployed by the framework authors to add other UI components to the framework. To get a feel for the gravity of the container's importance take a look at its class inheritance model in figure 4.1.

As illustrated in figure 4.1, the `Container` class is the base for at least 14 immediate subclasses, including widgets that you might not expect like `NestedList`, `Carousel`, and `Toolbar`.

In case you're wondering, the reason `Container` is used as the base for these widgets will become much clearer when we take a look under the hood of this widget.

4.1.1 Container's anatomy

The versatility of the `Container` class stems from its ability to contain child widgets and the ability to easily manage and arrange them. The container uses a layout manager to visually organize child items in an area known as the content body, or "body" for short. To understand what that looks like, go ahead and build your first container using the code in the next listing.

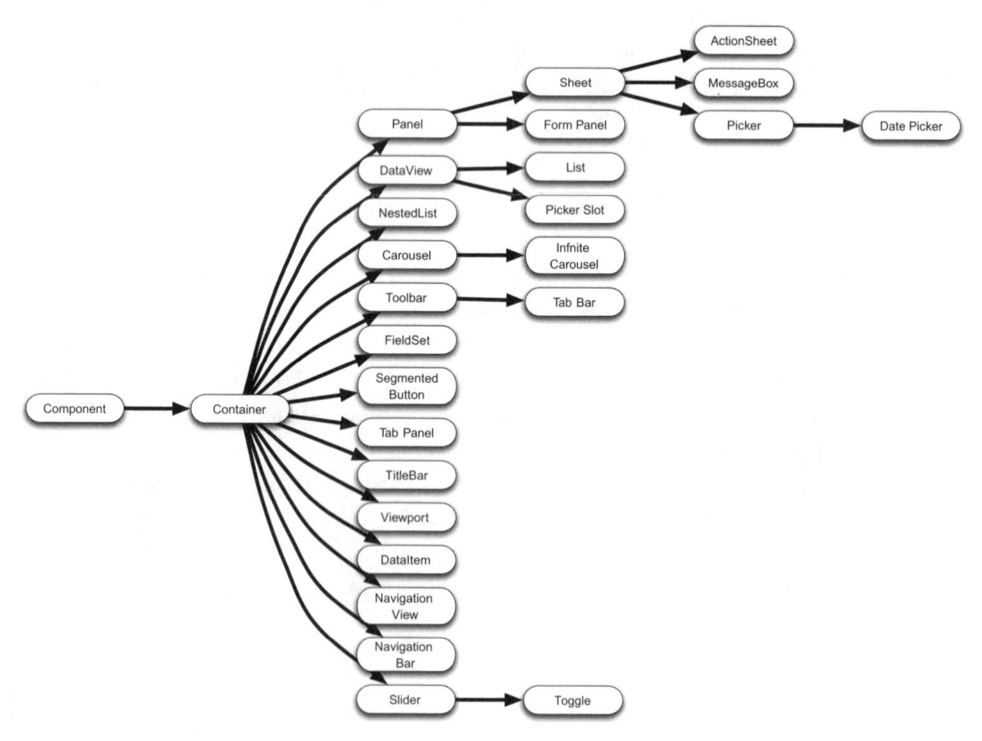

Figure 4.1 The `Container` class inheritance model

Listing 4.1 Building your first container

```
var loginContainer = Ext.create('Ext.Container', {
    itemId      : 'loginContainer',
    fullscreen : true,
    items       : [
        {
            xtype       : 'textfield',
            label       : 'Login',
            placeHolder : 'Enter Username Here'
        },
        {
            xtype       : 'textfield',
            label       : 'Organization',
            itemId      : 'orgField',
            placeHolder : 'Enter Your Organization Here'
        },
        {
            xtype       : 'textfield',
            label       : 'Password',
            placeHolder : 'Enter Password Here'
        }
    ]
});
```

The code itself is relatively simple. You instantiate a container and give it an `id`. Because you'll reuse this sample code in later examples assigning an `id` is necessary so you can obtain a handle on the container. This is your one and only container on the screen so you can mark it as `fullscreen: true`. Doing so ensures that the container automatically uses the full height and width of the device on which it's run. After that you add a few form elements via `XType`. Once the code executes it should look like figure 4.2.

Figure 4.2 The rendered container UI for listing 4.1

You have your first container, with a login form, no less. Obviously, you'll want to interact with the components for various reasons, such as obtaining values from the form elements, or because you need to change the content of the container to fit your particular application logic. To do so you'll need to learn how to obtain references to the container and its children and how to make changes to the container.

4.1.2 *Keeping unruly children on the right track*

Dealing with unruly children is a frustrating fact of life for any parent. Over the years you develop tricks that help with that endeavor. Remedies include timeouts, revoking TV privileges, or handing them off to your parents for a while.

Just as in real life, you must develop ways to handle children within containers to keep them in line. Luckily for you most of the helpers and utility methods you need to accomplish this are already in place. Mastering these methods will enable you to dynamically update the UI of your app and make it more interactive. Now let's get acquainted with these methods.

Adding components into a container is a relatively simple task, and you're provided with two methods to accomplish it: `add` and `insert`. The `add` method appends the new child component to the container's hierarchy, whereas the `insert` method places the new component at the specified index. To illustrate the `add` functionality further let's add to the container created in listing 4.1. To do this you'll use the handy browser debug console:

```
var container = Ext.ComponentQuery.query
     ('#loginContainer')[0];
container.add({
    xtype : 'checkboxfield',
    label : 'Remember'
});
```

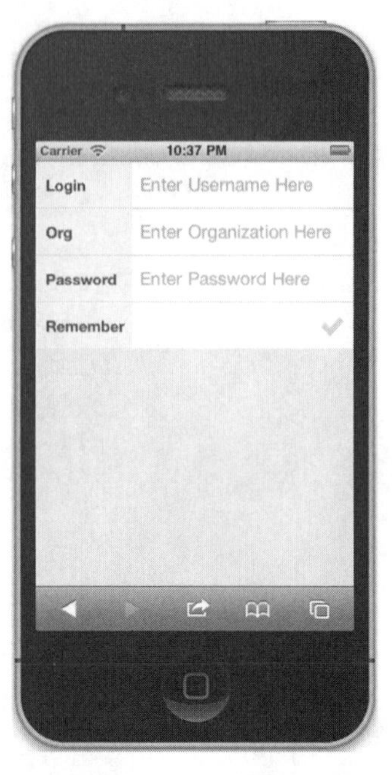

Figure 4.3 Illustration of code from listing 4.1 with a checkbox added dynamically; notice that the checkbox was appended at the end.

Running this code will add a new component (a checkbox) to the container (figure 4.3). The container's layout automatically handles refreshing the container and updating the position and size of all child components to make the new component appear in the right place.

> **Ext JS and Sencha Touch 1.x developers take note**
>
> For those developers upgrading from Sencha Touch 1.x (or coming from Ext JS), you might've noticed that the new component shows up immediately. This is due to the new layout manager which removes the need to call `doLayout()` separately.

Inserting a component works basically the same way as adding one; the main difference is the additional parameter that determines the index where the new component should be inserted. In this example you're inserting a new item at index 0, making the new component the first item in the container:

```
var container = Ext.ComponentQuery.query('#loginContainer')[0];
container.insert(0, {
   html: '<h1>Please enter your credentials</h1>'
   });
```

Running this snippet of code from the Safari debug console will produce what's shown in figure 4.4.

As you can see, adding and inserting components is easy as pie. Removing components is no different in that regard. You either provide a reference to the component that should be removed from the container or to the component's id or itemId:

```
var container = Ext.ComponentQuery.query
     ('#loginContainer')[0];
   container.remove('orgField');
```

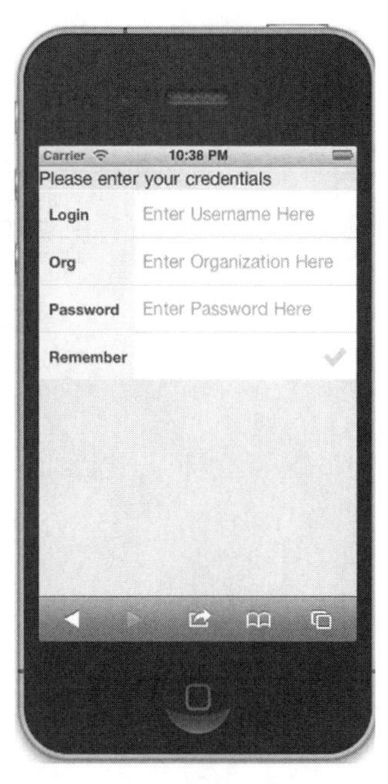

Figure 4.4 Rendered UI using code from listing 4.1 with dynamically added item shown in figure 4.3, as well as an inserted panel at position 0. Because the panel doesn't have any borders around it, it shows as an HTML snippet at the top.

Executing this statement from the debugger console causes the component to invoke its destroy method automatically, thus initiating the destruction phase and deleting the component's DOM element.

One special thing to note about the remove method is an optional second parameter that can be passed to tell the removed component to skip invoking the destroy method. Using this parameter is particularly useful if you wish to move a component from one container to another. In principle this works by obtaining a reference to the component you wish to remove and then removing the component while passing false as the additional second parameter. Doing so leaves the component on the screen and only removes it from the container. You then add the component (using the reference you still have) to the second container, thus triggering the DOM elements for the removed component being moved into the new container. To see this translated into code take a peek at the following listing.

Listing 4.2 Moving items between containers

```
var mainContainer = Ext.create('Ext.Container', {
    fullscreen : true,
    defaults   : {
        style : 'margin-bottom:30px'
    },
    items      : [
        {
            itemId : 'panel1',
            items  : [
                {
                    xtype       : 'textfield',
                    label       : 'Login',
                    itemId      : 'loginField',
                    placeHolder : 'Enter Username Here'
                },
                {
                    xtype       : 'textfield',
```

❶ Defines default values for children

```
                    label       : 'Password',
                    placeHolder : 'Enter Password Here'
                }
            ]
        },
        {
            itemId : 'panel2',
            items  : [
                {
                    xtype       : 'textfield',
                    label       : 'Organization',
                    placeHolder : 'Enter Your Organization Here'
                },
                {
                    xtype : 'checkboxfield',
                    label : 'Remember Login'
                }
            ]
        }
    ]
});

Ext.Function.defer(function() {                                    ❷ Defers function execution
    var myField = mainContainer.down('#loginField');

    var panel1 = mainContainer.down('#panel1');                    ❸ Obtains handle on children
    panel1.remove(myField, false);

    var panel2 = mainContainer.down('#panel2');
    panel2.add(myField);
}, 2000, this);
```

Listing 4.2 might look complicated at first but it breaks down relatively simply. You're creating an `Ext.Container` that contains two child panels, each of them with a different `itemId` and two form elements, `textfields` and `checkboxes` respectively. The main panel contains a `defaults` property ❶ that contains configuration options that are automatically applied to all of its direct children, in this case `panel1` and `panel2`.

After instantiation you create a function ❷ whose execution is deferred by 2 seconds. For this you're using the `Ext.defer` method, which is similar to the standard JavaScript `setTimeout` function. It defers the execution of a function by a specified number of milliseconds. What makes it special is that it allows you to set the scope of the function execution, thus enabling you to run the deferred function in the right context with the variables you need available. The deferred function first obtains a reference on the field you wish to move, the `loginField`, and then a reference to the panel you wish to remove it from, `panel1` ❸. You call `panel1.remove` and pass in the reference to the `loginfield` and `false` to indicate that you want to keep the `loginField` around after it's removed from `panel1`. You then obtain a handle on `panel2` and add the `loginField` there.

At this point you're probably wondering exactly how you obtained a reference to the items you wanted and how the `down` method works. For this and more stay tuned for the next part of the chapter.

4.1.3 Ask and ye shall receive: querying the container hierarchy

Containers provide multiple ways for you to find and get a handle on children. Generally these methods can be split into two groups. The first group contains methods that only return direct children of the container. The second group contains methods that return children at any level. To illustrate the difference, imagine an `Ext.Container` that contains an `Ext.Fieldset` with a single `textfield` element. The first set of methods would only be able to reach the `fieldset`, because that's a direct child of the panel. The second set of methods would be able to return the `textfield` because it can find items at any level by searching within items of the container.

Of all these methods, `getComponent` is by far the easiest to use, and it falls into the first group. It accepts only a single parameter which can either be an index or a string. The index corresponds to the index location of the item within the `items` collections of the container. The string corresponds either to the `id` or the `itemId` of the component you want to retrieve. In practice, using `getComponent` looks like this:

```
mainContainer.getComponent(0);
mainContainer.getComponent("panel2");
```

Not much to it, eh? The first line retrieves the first child component within `myPanel`, and the second line retrieves a component with an `id` or `itemId` of `panel2`. If you assume a code structure like that in listing 4.2, then the first line would retrieve `panel1` and the second line would retrieve `panel2`. Keep in mind that this method only returns direct children of the container you're querying.

Although `getComponent` is relatively simple to use its use is somewhat limited. This is why Sencha Touch presents you with three functions that have more flexibility in what and how you query.

All three of these functions take a single selector string that behaves similarly to CSS selectors, allowing you to query for items either by `id`, `itemId`, `XType`, or any combination thereof. To draw more parallels to CSS, you can search by `id` and `itemId` the same way as when you're addressing an `id` in CSS: via the pound sign. Likewise, `XTypes` are similar to classes, requiring the use of a dot to address them.

The first function is the `child` method, which falls into the first group we discussed earlier, returning only direct children of a container. You'll once again use the code structure from listing 4.2 as a baseline for the next example, which consists of multiple calls to the `child` method. Each call will pass in one of the selectors from table 4.1, just like this:

```
mainContainer.child("SELECTOR-GOES-HERE");
```

Looking at table 4.1, you can immediately see the benefits of using `child` to make your queries. As mentioned before, it isn't the only method. The other two are the `down` and `query` methods. Both follow the same syntax as the `child` method, with the only distinction being that all sublevels of a container are searched, not just direct children.

Table 4.1 Sample selectors that should help you differentiate between different selectors that the `child`, `down`, and `query` methods accept

Selector	Explanation and result
`"#panel1"`	The pound sign indicates that you're searching for an item by `id` or `itemId`, in this case, `"panel1"` (because `panel1` is the first item that's a direct child of `mainPanel` and matches the `itemId`).
`"#loginField"`	Here you're searching for the `loginField` by `itemId`. Because `loginField` isn't a direct child of `mainPanel` null is returned.
`".panel"`	This line searches for items with the panel `XType`. This is indicated by the dot in the beginning. This would return `panel1` because it's the first direct child with an `XType` of `panel`.
`".textfield"`	This line would once again return null because you're searching for items with an `XType` of `textfield`, none of which are direct children of `mainContainer`.

If you were to rewrite the second sample from table 4.1 to use `down` instead of `child` it'd look like the following and return the `login` field:

```
mainContainer.down("#loginField");
```

Both `down` and `child` return only a single item. More specifically, they return the first match encountered. If you anticipate having multiple results and want them all use the `query` method. Rewriting the last sample call from table 4.1 to use `query` returns an array of three items, all of them `textfields`:

```
mainContainer.query(".textfield");
```

The only thing to keep in mind about `query` that makes it slightly different from the other methods is that it always returns an array. Even if a single item or nothing was found it'd still return an array, albeit one that contained a single item or was empty.

A note about good form

When you're using the new `ComponentQuery` functions to search by `XType` a dot prefix (`.xtype`) is optional, but it's considered good form to use it anyway because doing so makes it clear that you're searching for an `XType`.

We've covered containers and how to fill them with components, as well as how to add, edit, and remove components from containers. Each example so far has only dealt with the most basic way to arrange components: stacking them on top of each other using the `default` layout. We're not sure about you, but we want our applications to be more visually appealing than simply having elements stacked one above the other. So let's take a closer look at the magic behind arranging components within your containers: the Sencha layouts.

Learn more about ComponentQuery

One little-known secret about the container's `child`, `down`, and `query` methods is that they implement something called `ComponentQuery` under the hood. `ComponentQuery` is a way to search your application for instances of components using Sencha's query language. The methods we just mentioned scope the query to a specific container instance, but you could search an entire application using `Ext.ComponentQuery.query()`. To learn more about this check out `Ext.ComponentQuery` in the Sencha Touch API documentation.

4.2 Everything must have its place: layouts

One of the hardest parts when building any app is managing all the UI pieces. Positioning everything correctly so it shows up where it's supposed to and making sure things remain that way after the user starts wildly thrashing around in your beautifully laid-out application are difficult to achieve. This is where the Sencha Touch layout management schemes come to the rescue. They're responsible for managing the visual organization of all components onscreen. The complexity of these layouts ranges from a simple fit layout, which automatically sizes a single child item to fit within the confines of its parent, to more complex layouts like the card layout, which provides a wizard-like interface with multiple screens.

When exploring these layouts we'll hit upon some more lengthy and verbose examples that can serve as a great starting point for your application. Let's start our journey by taking a look at the building block for all layouts, the default layout, shown in figure 4.5.

4.2.1 The default layout

The default layout serves as the foundation for all other layouts. Besides managing items that need to be laid out and positioning them it provides the capability to dock items. This ability is in turn passed down to all other layouts. You'll learn more in the next chapter.

The default layout is the easiest to use because it's automatically assigned for any container that doesn't have a specific layout defined. Any items within a container

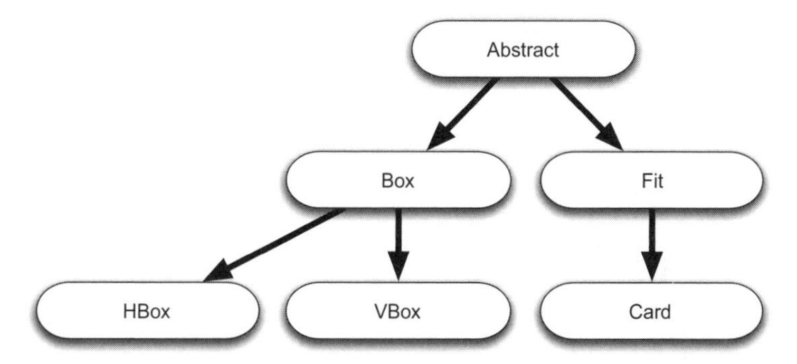

Figure 4.5 The `Layout` class hierarchy

using this layout are placed on the screen, one on top of another. Items aren't directly size-managed, meaning that any item can conform to the size of the parent, but it doesn't necessarily have to do so. To see this you need to set up a dynamic example using a few components, as shown in the following listing.

Listing 4.3 The default layout

```
var myContainer = Ext.create('Ext.Container', {
    fullscreen : true,                                   ❶ Use fullscreen option
    defaults   : {
        style : 'border: 1px solid blue;'                ❷ Contains default values for children
    },
    items      : [
        {
            docked : 'top',
            xtype  : 'toolbar',                          ❸ Creates title bar
            title  : 'Default Layout'
        },
        {
            docked : 'bottom',
            xtype  : 'toolbar',                          ❹ Creates bottom-docked toolbar
            items  : [
                {
                    text    : 'Add Child',
                    handler : function() {
                        myContainer.add({
                            xtype : 'container',
                            style : 'border: 1px solid blue;',
                            html  : 'Child'
                        });
                    }
                },
                {
                    text    : 'Add Fixed Width Child',
                    handler : function() {
                        myContainer.add({
                            xtype : 'container',
                            style : 'border: 1px solid blue;',
                            width : 100,
                            html  : 'Fixed Child'
                        });
                    }
                }
            ]
        },
        {
            html : 'First Child'                         ❺ Adds child panels
        },
        {
            html  : 'Fixed Width Child',
            width : 100
        },
        {
            html : 'Child'
```

```
        }
    ]
});
```

Listing 4.3 does quite a lot to exercise the default layout. The primary reason is to illustrate how items stack and don't resize when added dynamically.

The first thing you do is instantiate a container that occupies the entire screen through the `fullscreen` option ❶. Because you'll be adding simple child containers, which don't come with any visual indicators to mark their bounding boxes, you'll once again use the `defaults` object ❷ to automatically give each child a blue, one-pixel border. To be a bit more user-friendly you add a title bar ❸, which is a `toolbar` with `docked:'top'` and a `title` property. After that you add a second toolbar at the bottom, with two buttons ❹, to allow the addition of new items that have a fixed width and items that don't have a fixed width. Note that although the `toolbars` are defined inside of the `items` array, they contain the `docked` option, which means they don't show as part of the container body itself. The toolbar buttons have handlers that add a new child container into the main container. Last, you set up three static `container` instances in the `items` array property ❺. Notice that you aren't specifying any `XTypes` for the child items. Unless otherwise specified all direct children of a container use `"container"` as the default `XType`. When you run the sample you should get a screen like the one on the left side of figure 4.6. Be sure to hit the buttons in the toolbar a few times to see how items are sized once they're added, just like the one you see in the right side of figure 4.6.

> **A change from Sencha Touch 1.x**
>
> If you're upgrading from Sencha Touch 1.x you might be wondering what happened to `AutoContainerLayout` and `AutoComponentLayout`. The answer is simple: they've both been replaced by the default layout. They have, however, been added as alternate classnames for the default layout to provide backward compatibility, ensuring you don't have to change your existing apps to accommodate this change.

Although the default layout provides little to manage the size of child items, it isn't completely useless. Relative to the other layouts it's lightweight, making it ideal if you want to display child items that have fixed dimensions. There are times, though, when you'll want to have child items dynamically resize to fit the parent container. For those times the fit layout is exactly what you need.

4.2.2 *Make it fit: the fit layout*

The fit layout, shown in listing 4.4, forces a container's single child to "fit" to its body element and is by far the simplest of the layouts to use. It only requires that you have a single item within the container.

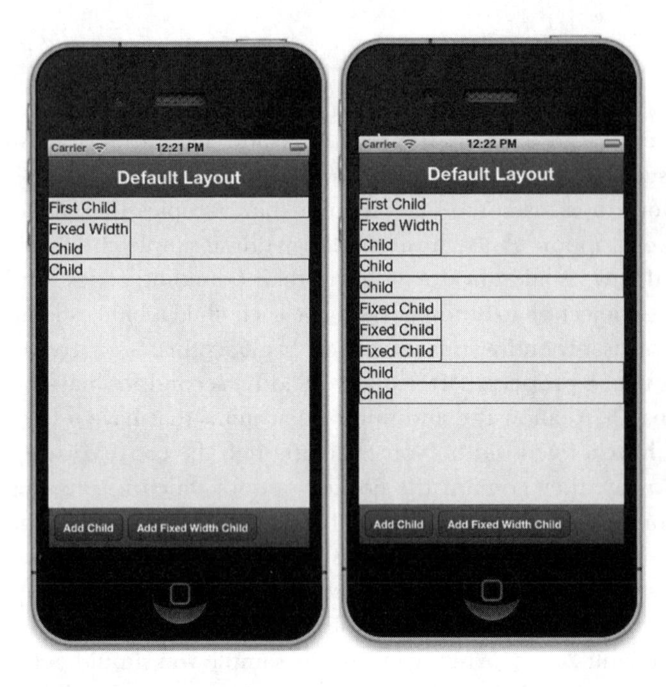

Figure 4.6 The result of a simple default layout. The left side shows the initial state of the code from listing 4.3, whereas the right side shows the result of clicking the Add Child and Add Fixed Width Child buttons a few times. The blue border makes it easier to see the bounding boxes for each item.

Listing 4.4 The fit layout

```
var myContainer = new Ext.Container({
    fullscreen : true,
    layout     : 'fit',
    items      : [
        {
            docked : 'top',
            xtype  : 'toolbar',
            title  : 'Fit Layout'
        },
        {
            xtype : 'container',
            style : 'background-color: pink; padding: 20px;',
            html  : 'I Fit in my parent'
        }
    ]
});
```

Listing 4.4 generates a simple panel that automatically occupies the entire screen of the device it's run on, as shown in figure 4.7.

One caveat to keep in mind about the fit layout is that it's only suitable if a single item is present. The layout will break if multiple items are housed within the same container, thus causing undesirable effects. For managing multiple items within a container take a look at the card and box layouts next.

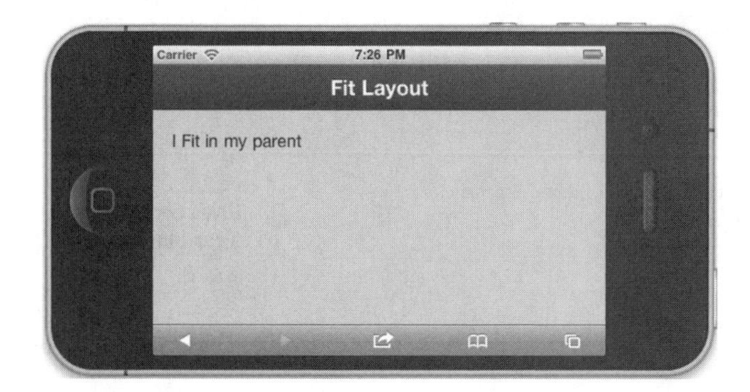

Figure 4.7 Using the fit layout to illustrate a single child resizing to fit its parent

4.2.3 *Card layout*

A direct subclass of the fit layout, the card layout combines the concept of components conforming to the size of the container from the fit layout with the ability to have multiple children under one container. The layout behaves somewhat like a carousel or wizard, only showing a single "card" or "screen" at a time. Each card automatically occupies the full size of the container and can house any number of `Component` instances. Managing which card is shown is done during render time using a container's `activeItem` configuration, and after rendering flipping between cards is left to the developer via the container's `setActiveItem` method. Under the hood, the card layout taps into the container's `activeitemchange` event to catch whenever it's supposed to change items and perform the change.

Navigation frequently occurs in a wizard-like interface through the use of previous and next buttons in a toolbar, or through actions embedded within the cards. Of course, there are various other methods to achieve the same thing: a programmatic timer or event-based methods, for example. In the next listing you explore how to create a simple wizard-style interface.

Listing 4.5 The card layout

```
var handleNavigation = function(btn) {                          // 1 Declares function to handle navigation
    var currentContainer = myContainer.getActiveItem(),         // 2 Retrieves currently active card
        innerItems       = myContainer.getInnerItems(),
        totalItems       = innerItems.length,
        currentIndex     = innerItems.indexOf(currentContainer),
        direction,
        newIndex;

    if (btn.getText() == 'Back') {                              // 3 Determines which button was pressed
        direction = 'right';
        newIndex  = currentIndex > 0
                        ? (currentIndex - 1) : (totalItems - 1);
    }
    else {
        direction = 'left';
        newIndex  = currentIndex < (totalItems - 1)
                        ? (currentIndex + 1) : 0;
    }
```

```
        myContainer.animateActiveItem(newIndex, {
            type      : 'slide',
            direction : direction
        });
    };

    var myContainer = Ext.create('Ext.Container',{
        fullscreen : true,
        activeItem : 1,
        layout     : {
            type      : 'card',
            animation : 'slide'
        },
        items      : [
            {
                xtype  : 'toolbar',
                docked : 'top',
                title  : 'Card Layout',
                items  : [
                    {
                        text    : 'Back',
                        ui      : 'back',
                        handler : handleNavigation
                    },
                    { xtype : 'spacer' },
                    {
                        text    : 'Forward',
                        ui      : 'forward',
                        handler : handleNavigation
                    }
                ]
            },
            {
                html  : 'Card 1',
                style : 'background-color: #99F;'
            },
            {
                html  : 'Card 2',
                style : 'background-color: #F99;'
            },
            {
                html  : 'Card 3',
                style : 'background-color: #9F9;'
            }
        ]
    });
```

❹ Gives container a card layout

❺ Sets active item

❻ Attaches nav handler to buttons

The first thing you do is create a method to control the card flipping ❶ by determining the active item's index and then calculating which index to show next. This is accomplished by first retrieving the `activeItem` using the `getActiveItem` method ❷, which returns a reference to the currently active container. From there you determine where in the set of cards the `activeItem` is, because the `animateActiveItem` method that allows you to navigate between different cards works based on the index of the new card to which you wish to move. To obtain the index use the standard array function indexOf, feeding it a reference to the activeItem. This tells you the location of the card within the items array of the container. By checking the text of the button that was pressed you can determine which direction to navigate ❸, either back or forward. The only thing left then is incrementing or decrementing the index to reflect the direction you're moving. Because you want to wrap around from the first card to the last, and from the last card to the first, you need to perform an additional check to ensure that the new index doesn't go below 0 or above the number of items you have in the container. If you don't want to wrap around you could change this same logic to skip the navigation when Previous or Next is clicked from the first or last card, respectively.

Now that you have a way to navigate between different cards in a card layout, you need to create said cards. To do this, start with a container, giving it a layout of "card" ❹. This automatically triggers certain changes in the container to make it listen to the activeItem config option and expose the getActiveItem and animateActiveItem methods. Use the activeItem config ❺ to start with the second card by setting the config option to 1. This works because cards are in a 0-based index, so index number 1 is actually the second item. After that, add a toolbar with two buttons and point the handlers of the buttons to the handy navigation function ❻ you created earlier. The only thing left to get what's shown in figure 4.8 is to add the cards as child items. For this you only need to specify the html property, because all child items are containers by default.

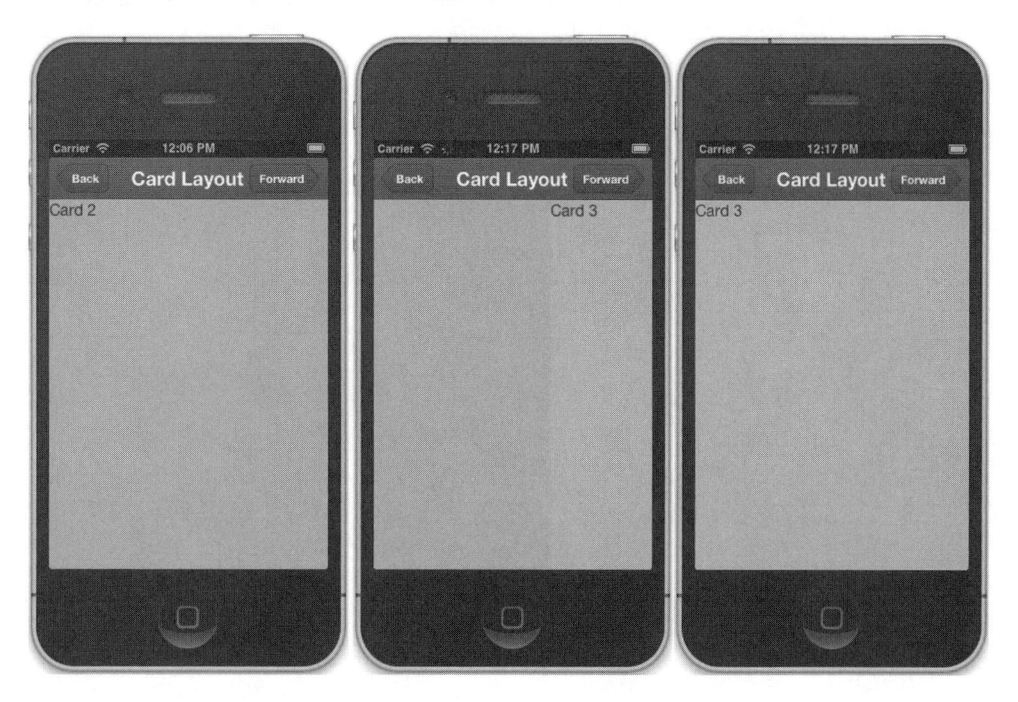

Figure 4.8 A card layout showing the initial card, along with a "cube" transition

The card layout is one of the most useful layouts because it provides a way to show content without much clutter, allowing the user to navigate through it one piece at a time. Be careful using it, though. Each card takes up valuable rendering time and memory. This is especially true on mobile devices where CPU cycles and RAM aren't in abundance. Adding too many cards can result in a significant slowdown of your application. The exact number when this occurs depends on the content of each card, and that's something you'll have to play around with. It's certainly something to keep in mind when designing your application.

Another major drawback of the card layout is that each card automatically occupies the entire container, thus eating up valuable screen space. To alleviate that you have another two layouts at your disposal that allow you to place multiple components on a screen in rows or columns: the box layouts.

4.2.4 HBox and VBox layouts

The `HBox` and `VBox` layouts are different from the ones we've discussed so far in that they display items in columns or rows respectively. This allows for much greater flexibility when creating complex layouts for your application, as each row or column could utilize its own layout, allowing you to combine a set of rows with a card layout in each row, for example. Let's dive into the `HBox` layout in the next listing.

Listing 4.6 The `HBox` layout

```
var myContainer = Ext.create('Ext.Container', {
    fullscreen : true,
    layout     : {
        type  : 'hbox',
        pack  : 'start',
        align : 'start'
    },
    defaults   : {
        style : 'border: 1px solid red;'
    },
    items: [
        {
            html   : 'Panel 1',
            height : 100
        },
        {
            html   : 'Panel 2',
            height : 75,
            width  : 100
        },
        {
            html   : 'Panel 3',
            height : 200
        }
    ]
});
```

In listing 4.6 you create a container with three irregularly shaped child containers to properly exercise the different layout configuration parameters available. The first is `pack`, or horizontal alignment, and the second is `align`, or vertical alignment. You'll notice that the sample code has a `defaults` property that sets a red border style, which will be automatically applied to all child items. This is done so you can better see the boundaries for each item, because containers don't have any borders by default.

The `pack` parameter accepts four different values: `'start'`, `'center'`, `'end'`, and `'justify'`. The first three options align the content on the horizontal axis either on the left, in the middle, or on the right side of the container. The fourth option, `'justify'`, aligns the content with the left *and* right side of the container at the same time, thus spreading content across the entire width. Modifying the `pack` parameter in listing 4.6 will result in one of the rendered panels in figure 4.9.

The `align` parameter accepts four possible values: `'start'`, `'center'`, `'end'`, and `'stretch'`. The first three options align at the top, middle, and bottom respectively, and the fourth option overwrites the height, thus stretching the content to occupy the entire height of the container. The default value for `align` is `'center'`. Figure 4.10 illustrates how you can change the way children are sized and arranged based on a few different combinations.

One of the great features behind box layouts is the ability to size rows and columns either manually, by specifying a fixed height and width as you did in listing 4.6, or dynamically via the use of a percentage value or a special `flex` value. The `flex` value acts as a weight, or a priority, if you will, instead of a percentage that a component should occupy. Let's assume for a second that you want to create a layout of multiple columns where all columns have an equal width. You would use an `HBox` layout and give each column the same `flex` value. If you wanted to have two of the columns

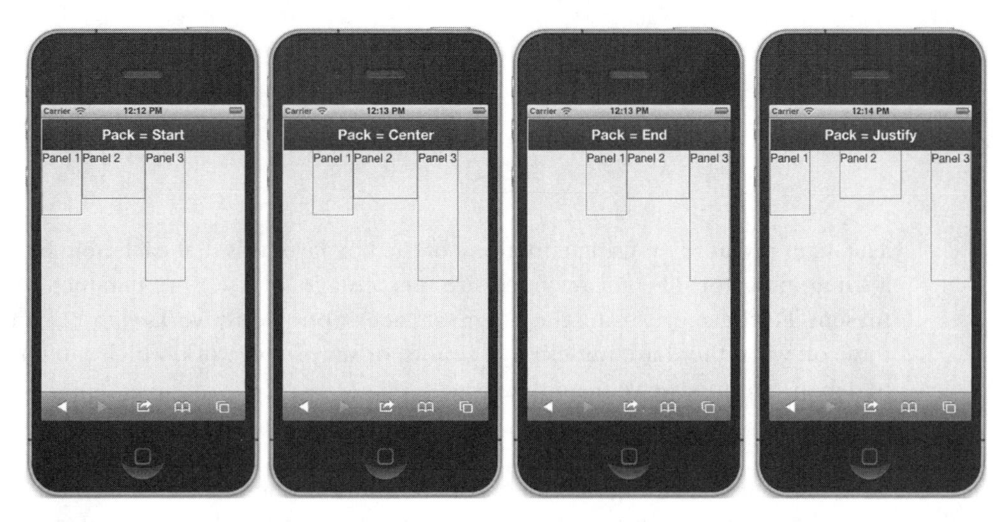

Figure 4.9 Showcasing the various `pack` options. These screenshots use an `align:start` setting as well.

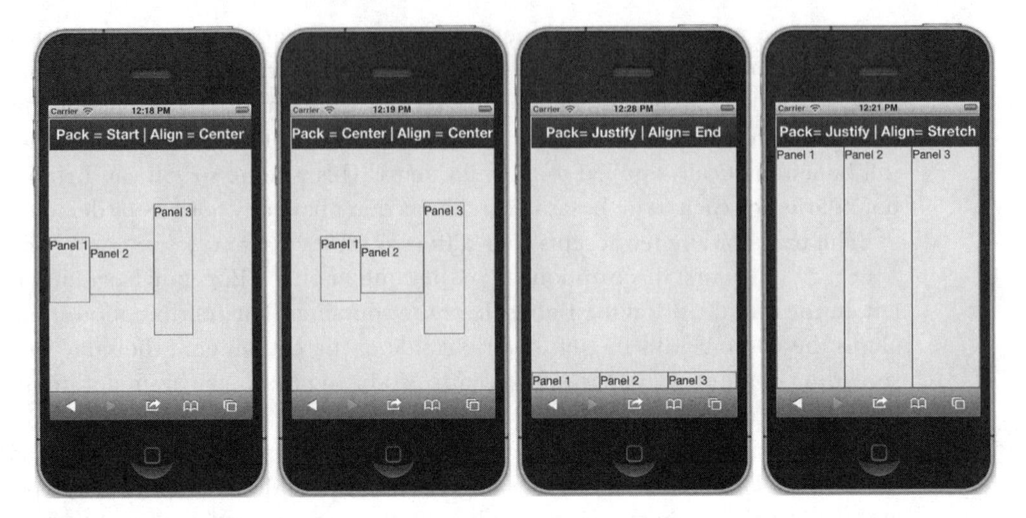

Figure 4.10 Mixing the `pack` parameter with different `align` options can produce interesting and useful combinations.

occupy the first half of the screen, and the third column expand to fill the other half, you'd have to ensure that the `flex` value for each of the first two columns is exactly half that of the third column. For instance:

```
items: [
    {
        html : 'Panel 1',
        flex : 1
    },
    {
        html : 'Panel 2',
        flex : 1
    },
    {
        html : 'Panel 3',
        flex : 2
    }
]
```

The main caveat when using either one of the box layouts is that each item has to have a sizing indicator. This means fixed size, percentage, or the `flex` parameter must be present. Furthermore, using the `align:stretch` option only works with `flex`. Leaving these off will either have undesirable results, or simply not work, which can take hours to debug. Trust us, we've been there.

4.2.5 Nesting layouts

Something many starting developers don't realize is that you can easily nest different layouts to achieve crazy combinations that don't generally exist out of the box. Because Sencha Touch doesn't provide for a table layout or column layout like Ext JS does, `HBox` and `VBox` are the only options at your disposal to achieve a multirow or multicolumn layout. Figure 4.11 illustrates what you can achieve with a little creativity, by nesting the `HBox` and `VBox` layouts and adding a modified version from the card layout sample covered earlier in listing 4.5.

Your goal is to create a grid-like structure with a carousel at the top that spans the full width, and two rows of content with three columns each, just like figure 4.11.

To get started you take a standard `Container` instance and give it a `VBox` layout. To achieve evenly spaced rows you specify a `flex` of 1 in the `defaults` property, thus forcing the same height on each row. Notice the additional `width` parameter that forces each direct child to have a width of 100%. This is needed because each child of the main `Panel` doesn't automatically adjust the width of the child items, and the `flex` parameter only applies to the height since you're using a `VBox` layout. Assembled, the code would look like this:

```
var mainContainer = Ext.create('Ext.Container', {
    fullscreen : true,
     layout    : 'vbox',
     defaults  : {
             width : '100%',
             flex  : 1
     },
     items : [
             // … content items go here …
     ]
});
```

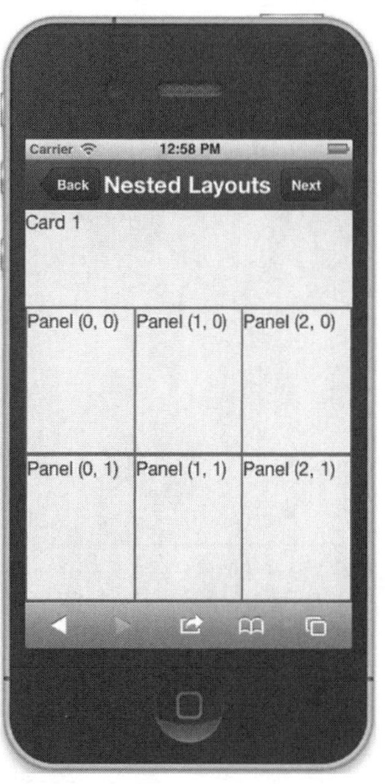

Figure 4.11 It's possible to nest different layouts to achieve more complex results. This could easily turn into a magazine layout, where the cards represent a picture gallery and the individual panels represent an article.

Now that you have your base container, you start adding content to it. See the following listing.

Listing 4.7 Nested layouts

```
var myContainer = Ext.create('Ext.Container', {
    fullscreen : true,
    layout     : 'vbox',
    defaults   : {
        width : '100%',
        flex  : 1
    },
    items      : [
        {
            itemId      : 'cardContainer',
            layout      : {
                type      : 'card',
                animation : 'slide'
            },
            activeItem : 0,
```

1 Obtains handle

```
                items       : [
                    {
                        xtype   : 'toolbar',
                        docked : 'top',
                        title   : 'Nested Layouts',
                        items   : [
                            {
                                text     : 'Back',
                                ui       : 'back',
                                handler : handleNavigation
                            },
                            { xtype : 'spacer' },
                            {
                                text     : 'Next',
                                ui       : 'forward',
                                handler : handleNavigation
                            }
                        ]
                    },
                    { html : 'Card 1' },
                    { html : 'Card 2' },
                    { html : 'Card 3' }
                ]
            },
            {
                layout    : 'hbox',
                style     : 'border: 1px solid blue;',
                defaults : {
                    style : 'border: 1px solid red;',
                    flex  : 1
                },
                items     : [
                    { html : 'Panel (0, 0)' },
                    { html : 'Panel (1, 0)' },
                    { html : 'Panel (2, 0)' }
                ]
            },
            {
                layout    : 'hbox',
                style     : 'border: 1px solid blue;',
                defaults : {
                    style : 'border: 1px solid red;',
                    flex  : 1
                },
                items     : [
                    { html : 'Panel (0, 1)' },
                    { html : 'Panel (1, 1)' },
                    { html : 'Panel (2, 1)' }
                ]
            }
        ]
    });
```

2 Creates panels with HBox layout

3 Provides HTML content

First up is the card layout code from listing 4.5, which you add into the `items` array. The only tweak you have to make to it is to give it an `itemId` **1** in listing 4.7 so you can obtain a handle on it later when you tweak the `handleNavigation` function. After that you add two `Panels` ❷ with an `HBox` layout and three child `Panels` ❸ each. You give each of the `Panels` a one-pixel border to make it easier to see where each one starts and ends. The three child `Panels` within each `HBox` container you give a `flex` of 1 and a height of 100%, as well as some simple HTML content. The `flex` is needed so that the three `Panels` are the same width and height, occupying the full height available. This is exactly the same paradigm as before with the `VBox` layout, but the height and width are reversed.

To finish it all up, you still need to tweak your handy navigation function for the carousel toolbar to fit into your new sample. The major difference between this version of the function and the one you had in listing 4.5 is that the card layout is no longer the main container, but instead is nested within another container. This means that you need to obtain a reference to it when you want to interact with it. You accomplish this by using the `down` method in conjunction with the newly added `itemId`.

As with listing 4.5, place the following function above your container instance (`myContainer`):

```
var handleNavigation = function(btn) {
    var cardContainer    = myContainer.down('#cardContainer'),
            currentPanel = cardContainer.getActiveItem(),
            innerItems   = cardContainer.getInnerItems(),
            totalItems   = innerItems.length,
            currentIndex = innerItems.indexOf(currentPanel),
            direction,
            newIndex;

    if (btn.getText() == "Back") {
        direction = 'right';
        newIndex  = currentIndex > 0
                        ? (currentIndex - 1) : (totalItems - 1);
    }
    else {
        direction = 'left';
        newIndex  = currentIndex < (totalItems - 1)
                      ? (currentIndex + 1) : 0;
    }
    cardContainer.animateActiveItem(newIndex, {
        type      : 'slide',
        direction : direction
    });
};
```

In the end you have a complex layout that can easily be used as a springboard for an application. Potential uses include an online magazine or a picture gallery. The possibilities are almost endless.

Now that you've mastered containers and their layouts to some degree, it's time to look at how Sencha builds on containers and expands their functionality through two `Container` subclasses: `Panel` and `TabPanel`.

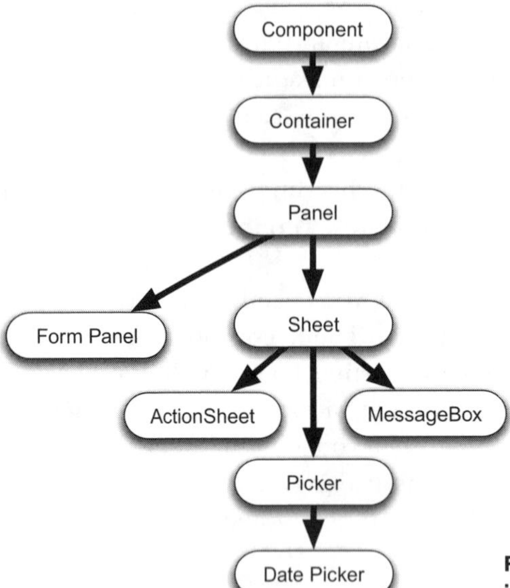

Figure 4.12 The `Panel` class inheritance model

4.3 Floating away... with panels

Figure 4.13 Floating panel with `showBy` used

Back in the olden days, the `Panel` class used to be the workhorse widget of Sencha Touch. A panel was the de facto go-to component whenever you needed to manage multiple child items. These days, however, the panel's dominance has been usurped by the standard container, relegating the `Panel` class to a simple extension of the `Container` class with some added styling, like a border, and the ability to float. It's important to note that because `Panel` does extend `Container` it carries with it all of the same abilities that `Container` has, including the use of layouts, docking items, and managing children. Figure 4.12 shows a quick overview of the full `Panel` inheritance chain.

As illustrated in figure 4.12 the `Panel` class is the base for everything floating and modal, such as sheets, pickers, and message boxes. This is because panels make for great overlays; they have special handling for the `showBy` method, where the panel shows a small tip pointing to the reference component that shows the panel. To see an illustration of this take a peek at figure 4.13.

Although figure 4.13 does look mighty pretty you're not here to simply look at pretty pictures. So let's get our hands dirty and implement a panel. Because we're touting the panel's ability to float, that's what we'll cover first. Let's up the ante, though, and add the ability to drag panels around as well while we're at it, as shown in the next listing.

Listing 4.8 Creating a draggable and floating panel

```
var floatingPanel = Ext.create('Ext.Panel', {       // 1 Creates floating panel
    height    : 200,
    width     : 200,
    draggable : true,                               // 2 Makes it draggable
    floating  : true,                               // 3 Makes it floating
    html      : 'Some help could go here.',
    left      : 50,                                 // 4 Moves to the 100 X, Y coordinates
    top       : 50,
    items     : [
        {
            xtype  : 'toolbar',
            docked : 'top',
            title  : 'Drag me!'
        }
    ]
});

Ext.Viewport.add(floatingPanel);                    // 5 Shows panel

var fsPanel = Ext.create('Ext.Panel', {
    fullscreen : true,
    style      : 'background-color: #CCF;',
    html       : 'Full screen Panel'
});
```

As shown in listing 4.8, to create a floating and draggable panel ❶ you must set the `draggable` ❷ and `floating` ❸ options to Boolean `true`. Setting `floating` to `true` instructs `Panel` to set its element as absolutely positioned on the page immediately after it has rendered on screen. Likewise, upon render `Panel` enables drag and drop by means of creating a new instance of `Ext.util.Draggable` for itself.

> **Know the root of this functionality**
>
> Even though we're using `Panel` to demonstrate floating and the drag and drop features, the root of this functionality is the `Ext.Component` class. This means that you can enable these features for just about any subclass of `Component`, which includes any custom widgets that you create. All of the features we'll be discussing moving forward will be from the perspective of `Panel`.

Next, you set the panel's position ❹ to 50 pixels in the X and Y coordinate space. Doing so demonstrates that the absolute positioning set forth by the `floating` configuration option works.

After creating and rendering the floating and draggable panel you need to add it to the automatically created `Ext.Viewport` instance ❺ in order to make the panel show up. This is because after the panel is created it isn't managed by any

Figure 4.14 Demonstrating the floating and draggable panel

parent container, and because it's floating you need to use the main viewport to make it visible.

Listing 4.8 concludes with the creation of a full-screen panel for contrasting purposes. You do this because the probability that you'll have a floating and draggable panel above a full screen and stationary panel is very high. Figure 4.14 illustrates how it renders on a phone.

After rendering your floating and draggable panel on screen you might notice that the floating panel looks more decorated than a nonfloating panel. This is because the `Panel` class wraps an extra thick border around itself when it's set to floating. It gives it more of a windowed UI feel.

Because you set `draggable` to true you can move the panel around the screen by means of a drag gesture. Notice how the panel is constrained to the limits of the display. This is a typical touch interface UI convention.

Postinstantiation floating and dragging

If you want to enable these features after a component has been instantiated you can use the `setFloating` and `setDragging` methods. The framework API has full details on how these methods work under the `Ext.Component` class.

Before we wrap up this discussion there are a few important nuggets of information you should know about floating panels.

The first is the fact that floating panels can be set to `modal`, meaning that Sencha Touch will create a `div` element underneath the floating panel, effectively creating a dimming effect on the rest of the screen. Doing so helps drive attention to your floating panel. To enable the modal feature, add the following configuration parameter to listing 4.8:

```
modal: true
```

Figure 4.15 has a side-by-side comparison of a modal floating panel and a nonmodal floating panel.

Figure 4.15 illustrates how a modal floating panel (left) has a dimming effect on the rest of the screen, making everything else but the floating panel look darker compared to the non-modal floating panel, which seems to have the same contrast as the rest of the UI.

The next tidbit you should know about floating panels is that they'll self-hide whenever the user taps outside of the confines of the panel. This is the default behavior of a modal panel, but you can modify it via the following parameter:

```
hideOnMaskTap : false
```

The `hideOnMaskTap` configuration property defaults to Boolean `true`, which enables the default behavior. To alter the default set it to `false`. Doing so will ensure that the `Panel` won't hide automatically when tapping outside of the panel's element. One thing to note is that this default behavior is enabled whether or not you set `modal` to `true`.

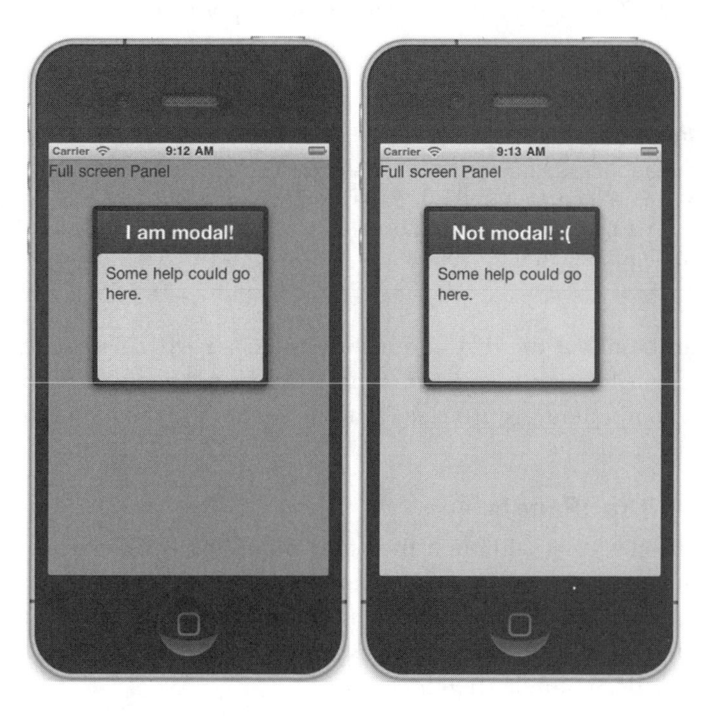

Figure 4.15 Comparing a modal floating panel (left) to a nonmodal floating panel (right)

Don't leave your users hanging!

We've run into situations where developers introduced bugs by disabling `hide-MaskOnTap` and not giving the users any way to dismiss the dialog. Remember, if you're going to disable `hideOnMaskTap` your users will have no way to get rid of the modal panel. This means you'll need to create a button that allows them to do so or hide the modal panel after a certain time.

Another feature to be aware of is that the appearance and disappearance of the panel can be animated relatively easily. All you have to do is set the following attributes:

```
showAnimation : 'pop',
hideAnimation : 'fade'
```

The `showAnimation` configuration property is used by `Panel` to animate the appearance of the panel, whereas `hideAnimation` is used to animate the disappearance. Each of these parameters accepts a `String` object. You would pass a Boolean `false` if you wanted to disable a component's default transition. For instance, you might do this if you created a new instance of `Ext.MessageBox` and didn't want to use its animation for `show` and `hide` actions.

Use a string where you want to use any one of the preconfigured animations from the `Ext.anims` singleton. The best ones to use are `pop` and `fade` because they're designed to transition in a single item, compared to the `cube` or `flip` animations which are designed to animate between two different components. If you only set a `showAnimation` Sencha Touch will automatically do the opposite animation for hiding the panel.

Animation performance is crucial

Be aware that animations use CSS3 transitions and work best on devices that have a dedicated GPU and are complemented with an instance of mobile WebKit compiled to utilize the GPU. Enabling animations on devices that can't handle them will result in a clunky UI and ultimately turn off your users.

We've covered a fair bit about the first of our two `Container` extensions, the Sencha Touch `Panel` class. You learned about `floating` and `dragging` and how to show a panel with another component. As promised, let's delve into the second of the extensions, the TabPanel.

4.4 Flip the deck with TabPanels

`TabPanel` is a special class that, although the name might imply differently, extends `Container`. The widget allows you to easily create an interface for users to navigate between screens by tapping on a special button known as a tab, located in a special toolbar called the TabBar. The TabPanel mimics the traditional desktop tabbed interface but within the modern mobile UI guidelines.

To fully understand how they work and how they can be customized you'll create a base example in the next listing that you can modify later on.

Listing 4.9 Creating a simple TabPanel

```
Ext.create('Ext.TabPanel', {                         ❶ Creates TabPanel
    fullscreen     : true,
    ui             : 'light',
    tabBarPosition : 'top',                          ❷ Configures TabPanel
    items          : [
        {                                            ❸ Adds TabPanel child items
            style : 'background-color: #FCC;',
            html  : 'Panel 1',
            title : 'Users'                          ❹ Sets title for tab
        },
        {
            style : 'background-color: #CFC;',
            html  : 'Panel 2',
            title : 'Admins'
        },
        {
            style : 'background-color: #CCF;',
            html  : 'Panel 3',
            title : 'Locations'
        }
    ]
});
```

Listing 4.9 represents an example of a simple TabPanel ❶ that contains four child panels ❸. You use `tabBarPosition` ❷ to set whether the TabBar should appear at the top or bottom of the TabPanel. In this case, you specifically defined it as `top`, but given that that's the default you could've just left it off. Lastly, you configure an array of `items` ❹ for the TabPanel to manage. These are generic configuration items that will be used to create instances of `Container`. What's important to focus on is the `title` property. This is not a normal property for `Container`. In this case, it'll end up being used to set the title text of instances of `Tab` that will live in the TabBar. Figure 4.16 shows what all of this looks like rendered onscreen.

Figure 4.16 illustrates the result from listing 4.9, which contains your simple TabPanel with left-aligned tabs. To display any of the screens the user can tap on any of the tabs, which will use the default slide animation transition.

Figure 4.16 The results of listing 4.9: a rendered TabPanel

If you wanted to change the alignment of the TabBar items to be right-aligned, for example, you'd have to utilize the `tabBar` option to override the default alignment. To do so, add the following code to listing 4.9:

```
tabBar : {
    layout : {
        pack : 'right'
    }
}
```

The `layout` parameter might look immediately familiar from earlier parts of the chapter. If so, you're absolutely correct, since the TabBar uses an `HBox` layout under the hood. So by adding `pack: "right"` you override a portion of the layout to align items differently.

The same paradigm would be true if you wanted to change the animation used when switching between cards. Let's say you wanted a `cube` animation instead of the default `slide`. In that case, add the following code to listing 4.9:

```
layout : {
    animation : 'cube'
},
```

The `animation` configuration property is part of the layout options for the TabPanel and instructs the TabPanel to animate the transition between the different cards (screens). You set this property to use the `cube` animation, which is different from the default `slide` animation. To disable animations you set this property to `false`. Generally, you want to consider disabling `animation` when transitions become sluggish.

> **See all of the animations**
>
> You're given the choice of `fade`, `slide`, `flip`, `cube`, `pop`, and `wipe` animations to use for `TabPanel` transitions. You can see all of these animations in the "Kitchen Sink" example, which can be found in the examples directory of the Sencha Touch SDK.

You might notice that when you top-align the TabBar the rendered tabs only show the given text and look rather plain. Given the way tabs are constructed from a class perspective, you can't easily add icons to them without an override, extension, or plug-in of some sort. This means you'll either have to go through quite a bit of pain to hack those in yourself, or bottom-dock the TabBar, which gives you the ability to set an `iconCls` property for each child item in the TabPanel. To better illustrate this, we've included an example in the following listing that modifies listing 4.9 to do just that.

Listing 4.10 Showing a bottom-docked TabBar for a TabPanel

```
Ext.create('Ext.TabPanel', {
    fullscreen : true,
    ui         : 'light',
```

❶ Docks TabBar to bottom

```
    tabBar      : {
        docked : 'bottom',
        layout : {
            pack : 'center'
        }
    },
    items       : [
        {
            style   : 'background-color: #FCC;',
            html    : 'Panel 1',
            title   : 'User',
            iconCls : 'user'
        },
        {
            style   : 'background-color: #CFC;',
            html    : 'Panel 2',
            title   : 'Groups',
            iconCls : 'team'
        },
        {
            style   : 'background-color: #CCF;',
            html    : 'Panel 3',
            title   : 'Locations',
            iconCls : 'maps'
        },
        {
            style   : 'background-color: #FFC;',
            html    : 'Panel 4',
            title   : 'Settings',
            iconCls : 'settings'
        }
    ]
});
```

To dock the TabBar to the bottom of a TabPanel you provide an alternative to the `tabBarPosition` you used before by injecting a `docked: bottom` configuration ❶ into the `tabBar` config, thus piggybacking on the fact that you're already using the `tabBar` config. Next, you add an `iconCls` property ❷ for the items in the TabPanel to dress them up a bit. Figure 4.17 illustrates the newly configured tabs rendered in the bottom-docked TabBar.

Figure 4.17 Your TabPanel with a bottom-docked TabBar containing tabs that display icons

As illustrated in figure 4.17, the bottom-docked TabBar renders with the tabs displaying the icons you configured for them. It's important to note that the `iconCls` property is required for tabs in bottom-docked TabBars. If you don't specify `iconCls` the tabs will render but will be barely usable as they'll be extremely small, just like in figure 4.18.

Figure 4.18 Bottom tabs without icons

There you have it! You've just seen how you can configure TabPanels with both top and bottom-docked TabBars. You should be able to use these lessons when developing your TabPanel-enabled application.

4.5 Summary

We've covered a lot of ground in this chapter. You learned how containers manage child items, and how to add and remove components dynamically in an already-rendered UI. You also learned how to arrange components in various ways using layout schemes. Although we covered enough material to make just about any developer's head spin, we barely scratched the surface of what's possible.

That being said, you may have noticed a recurring theme across many of the samples you encountered in this chapter. More specifically, you used toolbars and docked items in quite a few of them. Given the prevalence of toolbars and docked items in even the simple examples, it's important for you to better understand the capabilities of these powerful tools at your disposal. So in the next chapter you'll do exactly that by delving deeper into toolbars, buttons, and docked items.

5 Toolbars, buttons, and docked items

This chapter covers

- Unlocking the secrets of docking
- Mastering the toolbar
- Using and customizing buttons

In previous chapters we covered a lot of foundational topics, including the component life cycle, containers, and layouts. Many of the examples encountered thus far made use of docked items via toolbars, which set the stage for you to master the topics in this chapter, including toolbars, buttons, or anything else you want to dock.

You're already aware that containers and panels have the ability to dock widgets, so this chapter focuses on expanding on that concept. You'll learn how to dock just about any widget to any of a container's outer quadrants. In the process, you'll discover the importance of the order of docked items. You'll see a variety of ways to organize buttons and other widgets inside the toolbar. There you'll learn how to use the spacer component as well as customize the toolbar's `HBoxLayout` implementation.

The chapter ends by unraveling the secrets of buttons, and you'll learn how to customize their appearance by exploiting the built-in configuration options. Although this chapter is a bit shorter and doesn't cover quite as many topics as the others, the experience gained will be absolutely necessary for almost any application you build.

5.1 Looking into docked items

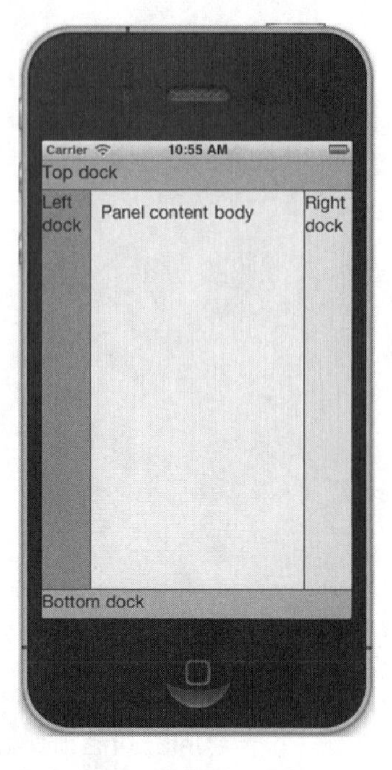

Figure 5.1 A simple panel example with four docked containers

It's relatively easy to create a container with items docked to a quadrant, but there are some hidden nuances that we've discovered along the way. To illustrate these we'll provide a basic example you can play around with.

5.1.1 Understanding the basics

As with most things in life, a picture is worth a thousand words. To get a better understanding of the basic example you're going to build, take a quick peek at figure 5.1, which shows a docked container on each outer quadrant of a panel.

This image illustrates docked containers at each possible outer quadrant of a panel. A panel (or any container, really) will allow any amount of docked items at any given time. For learning purposes you're going to include only four, one in each quadrant. The following listing shows the code that results in figure 5.1.

Listing 5.1 Placing containers at each of the outer panel quadrants

```
var topDock = {                                                    <-- 1 Configures docked item objects
        xtype  : 'container',
        docked : 'top',                                             <-- 2 Contains top-positioned docked item
        style  : 'border-bottom: 1px solid; background-color: #F99;',
        height : 100,
        html   : 'Top dock'
    },
    bottomDock = {
        xtype  : 'container',
        docked : 'bottom',
        style  : 'border-top: 1px solid; background-color: #9F9;',
        height : 100,
        html   : 'Bottom dock'
    },
    leftDock = {
        xtype  : 'container',
        docked : 'left',
        width  : 100,
        style  : 'border-right: 1px solid; background-color: #99F;',
        html   : 'Left dock'
    },
    rightDock = {
        xtype  : 'container',
        docked : 'right',
```

```
        width   : 100,
        style   : 'border-left: 1px solid; background-color: #FF9;',
        html    : 'Right dock'
    };

var myPanel = Ext.create('Ext.Panel',{
    fullscreen   : true,
    bodyStyle    : 'padding: 10px;',
    html         : 'Panel content body',
    items : [                                          <-- Registers docked
        topDock,                                       ③  items with Panel
        bottomDock,
        leftDock,
        rightDock
    ]
});
```

Listing 5.1 contains the necessary code to dock a container at each of the panel's outer quadrants. It begins with registering references to configuration objects ❶ for each of the docked containers.

> **Just about anything can be docked**
> You're using the container in listing 5.1 to reduce the code complexity of this example, but you can place pretty much any widget in a panel's outer quadrant.

In order to control where a widget is docked you must set a `docked` property ❷. There are four possible values: `top`, `bottom`, `left`, and `right`. Each value controls in which outer quadrant you want to render the widget.

Then you register the docked containers with the newly created panel by setting the panel's `items` property to an array ❸ containing references to each of the configuration objects you created earlier. Notice that even though the docked items are within the items array, they show up in their docked position and don't interfere with the panel content body.

> **Docking multiple items in a quadrant**
> Although our examples only dock a single item in each quadrant you could easily duplicate one of the configuration objects in the `items` array, and thus dock multiple items in one of the quadrants. The additional items would stack. Keep in mind that the order matters.

Now you have a good handle on how to register docked items. One thing that might be a bit confusing is that the order in which you register docked items will affect how the items are laid out in the quadrants. Take the configurations in figure 5.2, for instance.

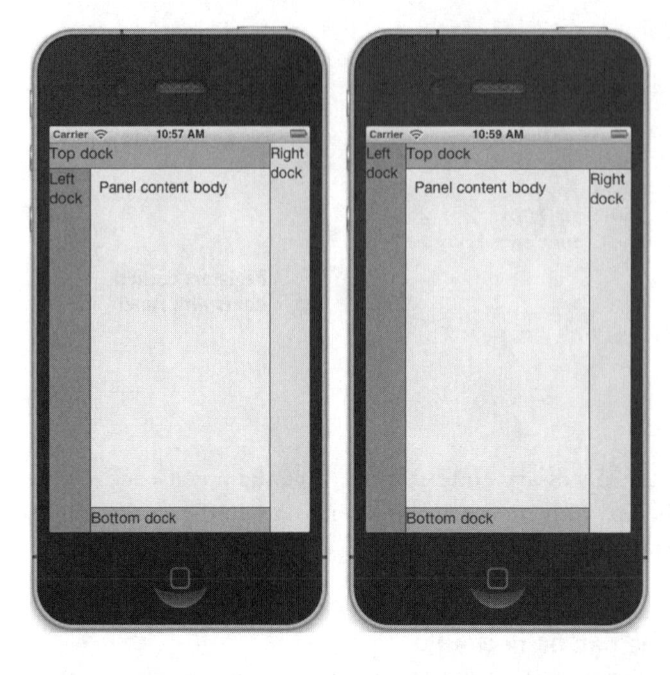

Figure 5.2 The order in which you place docked items will affect the amount of screen space they have.

Figure 5.2 demonstrates dock configurations that are different than the example you created in listing 5.1. This is because the order in which you list docked items controls how they're sized. The rule for sizing is simple: the docked items with the lower indexes get the largest portion of a quadrant.

To test this rule, you'll have to modify the `items` panel configuration property in listing 5.1 and shuffle the order of the docked `items` as follows:

```
items : [
    leftDock,
    rightDock,
    topDock,
    bottomDock
]
```

With the left- and right-docked `items` listed first, the docked containers would render as shown in figure 5.3.

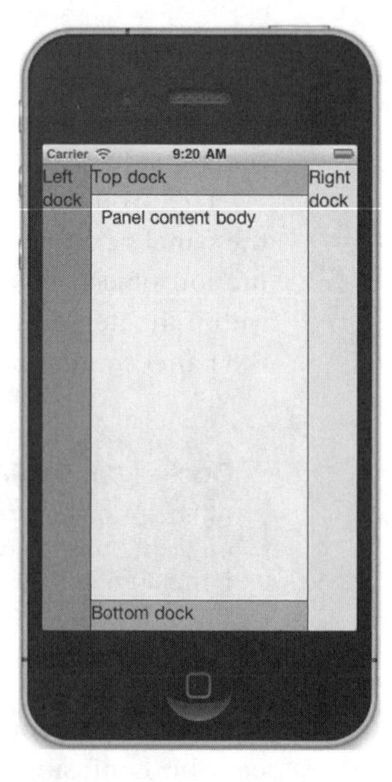

Figure 5.3 The result of your first modification to listing 5.1

As predicted, the left- and right-docked containers are larger than the ones located at the top and bottom. What if you wanted the left and right items to be the two that are the largest? You'd again have to modify the order in which you list the `items`. Given what you've just learned, can you predict how you should organize the `items` to achieve this desired result?

If you guessed the following order you're absolutely correct:

```
items : [
    leftDock,
    topDock,
    rightDock,
    bottomDock
]
```

The result of your second change to listing 5.1 renders as shown in figure 5.4.

As predicted, the left dock is the largest of the four and the top takes as much space as it can, whereas the right captures all available space in its quadrant. The bottom dock is the absolute smallest.

One secret to understanding how this all works is to know that the order of precedence rule only applies to docked items that are intersecting. For instance, the top intersects with the left- and right-docked but never the bottom. The bottom dock could care less about how large the top is. Likewise, the right dock intersects with the top and the bottom docks. The right doesn't fret about the left dock, nor does it care that the left dock is even rendered on screen.

Now you have a solid grasp on the docked items order of precedence rule, but you're not quite done with docked items yet. To fully realize their power, let's look at how you can use the dynamic nature of docked items to create dynamic UIs that can expand the workflows of your application.

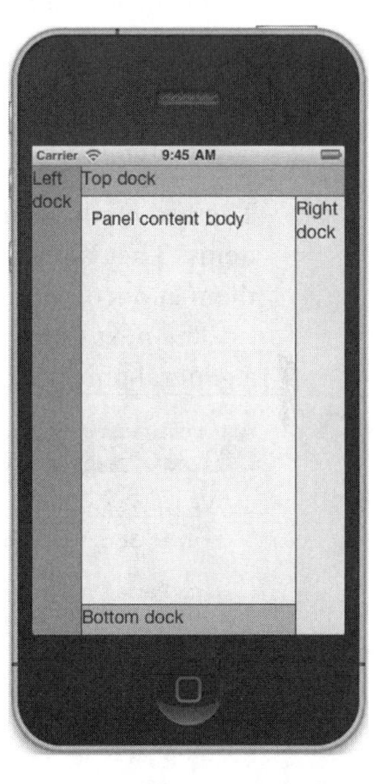

Figure 5.4 The result of your second modification to listing 5.1

5.1.2 Dynamic docking

In chapter 4 you learned how to dynamically add and remove items in a container via the `add` and `remove` methods. Adding docked items works exactly the same way. In fact, it uses the same methods (`add` and `remove`) as any other components would. The only difference is that docked items must have the `docked` property set. The `add` and `remove` methods will automatically figure out if the item is a docked item and do the appropriate thing.

> **For those developers upgrading from Sencha Touch 1.x**
>
> In previous iterations of the framework you had to employ the `addDocked` and `removeDocked` methods to make changes to docked items. With Sencha Touch 2.x these methods have been deprecated, and their functionality has been rolled into the standard `add`/`remove` methods.

Containers (and thus panels as well) use two convenient functions to get hold of a specific docked item or all docked items at once (see table 5.1).

Table 5.1 The methods used to manage docked items dynamically

Method	Description
`getDockedComponent`	This method is responsible for retrieving and returning a reference to a component that's in the docked collection. It takes a single argument, the `itemId` (`String`) of the component or the index (integer).
`getDockedItems`	This method returns a list of all currently docked items.

The two methods in table 5.1 are particularly useful if you need to work with docked items. They allow you to surgically target a specific docked item or work with all of them at once, without having to worry that nondocked items will interfere.

The next listing demonstrates adding and removing top-docked items by means of a general-purpose button rendered inside a panel.

Listing 5.2 Adding and removing `dockedItems`

```
var handleAddButton = function () {
    var dockedItems = myPanel.getDockedItems();

    myPanel.add({
        xtype  : 'container',
        docked : 'top',
        style  : 'border-bottom: 1px solid; background-color: #F99;',
        height : 30,
        html   : 'Top dock: ' + dockedItems.length
    });
};

var handleRemoveButton = function () {
    var dockedItems = myPanel.getDockedItems(),
        totalItems  = dockedItems.length,
        dockedItem;

    if (totalItems > 0) {
        dockedItem = myPanel.getDockedComponent(totalItems - 1);
        myPanel.remove(dockedItem, true);
    }
};

Ext.create('Ext.Panel', {
    fullscreen : true,
    bodyStyle  : 'padding: 10px;',
    layout     : {
        type : 'vbox',
        pack : 'center'
    },
    items      : [
        {
            xtype   : 'button',
```

❷ Handles button clicks

❸ Adds top-docked item

❹ Removes last top-docked item

❶ Creates panel

```
            text    : 'Add top',
            handler : handleAddButton
        },
        {
            xtype   : 'button',
            text    : 'Remove top',
            handler : handleRemoveButton
        }
    ]
});
```

Listing 5.2 contains the code to render a general-purpose panel with two buttons in the middle: one button to add a new top-docked item, and one button to remove the last top-docked item. Here's how you make it all work.

Begin by creating a panel ❶ with two buttons defined as simple child items within the panel. Use a `VBox` layout with `pack: center`, which will vertically align the buttons in the center. The first button is for adding top-docked items; the second is for removing docked items. Each of the buttons has a handler defined that points to a function to handle the tap event. The handler for the add button ❷ starts out by getting hold of all the docked items. You need to do this if you allow for adding multiple docked items, because you want to be user-friendly and allow the user to tell the different items apart. Do this by dynamically tweaking the configuration for the items you're adding ❸, and setting the `html` property of the new item based on the docked items count. The configuration object specifies `docked: top`, which causes the newly added item to be stacked in the top quadrant. To add the top-docked item you simply use the `add` method of the panel.

The `remove` handler also works by first getting all of the docked items. It then checks to see if any exist, and if so, retrieves the last docked item. Because the docked items use an array, which is a 0-based index, the length of the array will always be one more than the index of the last item; hence the need to use `length-1`. Use the `getDockedComponent` method in conjunction with the index to retrieve the last docked item, which is then removed by passing the item to the `remove` method ❹ of the panel. Notice the additional Boolean value passed to the `remove` method. This is to ensure that the docked item is destroyed and not just removed from the panel. In case this doesn't ring a bell, a revisit to chapter 4 might be necessary. Figure 5.5 shows the example rendered onscreen.

In figure 5.5 you can see that a panel renders with two buttons in the center of the screen (left). A couple taps of the button add and dock a container to the top dock, demonstrating the `add` method in conjunction with the `docked` config property. Any subsequent taps on the remove button will remove one of the top-docked containers, which demonstrates the `remove` function.

You've successfully demonstrated how to add, remove, and query for docked items. Next up, we'll elaborate on one of the most commonly docked items, the toolbar.

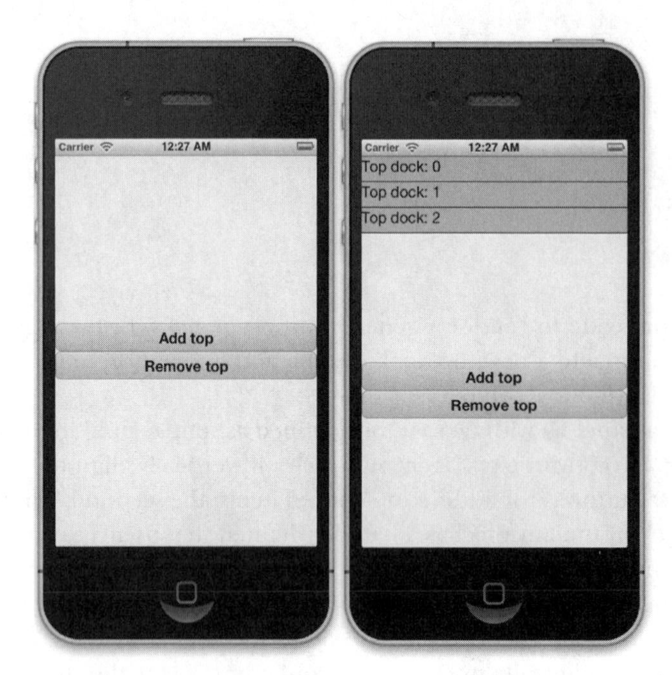

Figure 5.5 The initial results of listing 5.2 (left), and the same after adding the top docked container by tapping the button a few times (right)

5.2 Gearing up the toolbars

`Toolbar` extends the `Container` class. The toolbar is typically docked at the top or bottom, and it's used to house a set of buttons and text elements, such as a centered title. The toolbar is more versatile than that and in fact can be docked anywhere and contain pretty much anything, not just buttons and text.

5.2.1 Under the hood

The secret to `Toolbar`'s versatility is that it's a subclass of `Container`, which means that it can render and manage just about any component in the Sencha Touch framework. Before you get ahead of yourself, take a step back and create a basic example you can build on in the next listing.

Listing 5.3 Constructing a simple toolbar

```
var toolbar = {
    xtype   : 'toolbar',
    docked  : 'top',
    title   : 'User admin'
};

Ext.create('Ext.Panel', {
    fullscreen    : true,
    style         : 'background-color: #CCF;',
    html          : 'Full screen Panel',
    items         : toolbar
});
```

❶ Configures toolbar

❷ Sets title

❸ Includes toolbar

Figure 5.6 A simple toolbar with a title, rendered on a tablet

Listing 5.3 contains a recipe for a panel that includes a top-docked toolbar ❶. To keep things simple for now, you give the toolbar a title ❷ and nothing more. Because you're going to be building on this listing you'll use a tablet to view it; that will give you the most amount of screen real estate to work with. Figure 5.6 shows how your freshly minted toolbar ❸ looks when rendered on a tablet.

Figure 5.6 shows you how something as simple as adding a toolbar with a title can easily dress up an application. Though there are many situations where having a toolbar with just a title is sufficient, more often than not you'll run into situations where you need buttons for users to interact with as well. This is where things get interesting, because toolbars can contain a mixture of buttons, text, and much more.

5.2.2 Adding buttons to a toolbar

The following listing contains a modification of listing 5.3, where you redefine the `Toolbar` configuration object by adding two buttons and a spacer component.

Listing 5.4 Adding buttons to your simple toolbar

```
var toolbar = {
    xtype  : 'toolbar',
    docked : 'top',
    title  : 'User admin',
    items  : [
```

```
            {
                xtype : 'button',          <-- 1 Configures simple button
                text  : 'Submit'
            },
            {
                xtype : 'spacer'           <-- 2 Adds spacer
            },
            {
                xtype : 'button',
                text  : 'Cancel'
            }
        ]
};

Ext.create('Ext.Panel', {
    fullscreen  : true,
    style       : 'background-color: #CCF;',
    html        : 'Full screen Panel',
    items       : toolbar
});
```

Listing 5.4 contains a toolbar with a title, two buttons ❶, and a spacer component ❷. To drill down into what all of this is doing under the covers you must first see how it's rendered onscreen, as shown in figure 5.7.

Figure 5.7 illustrates how the newly modified top-docked toolbar renders on screen. Notice how the submit and cancel buttons are on opposite ends of the screen.

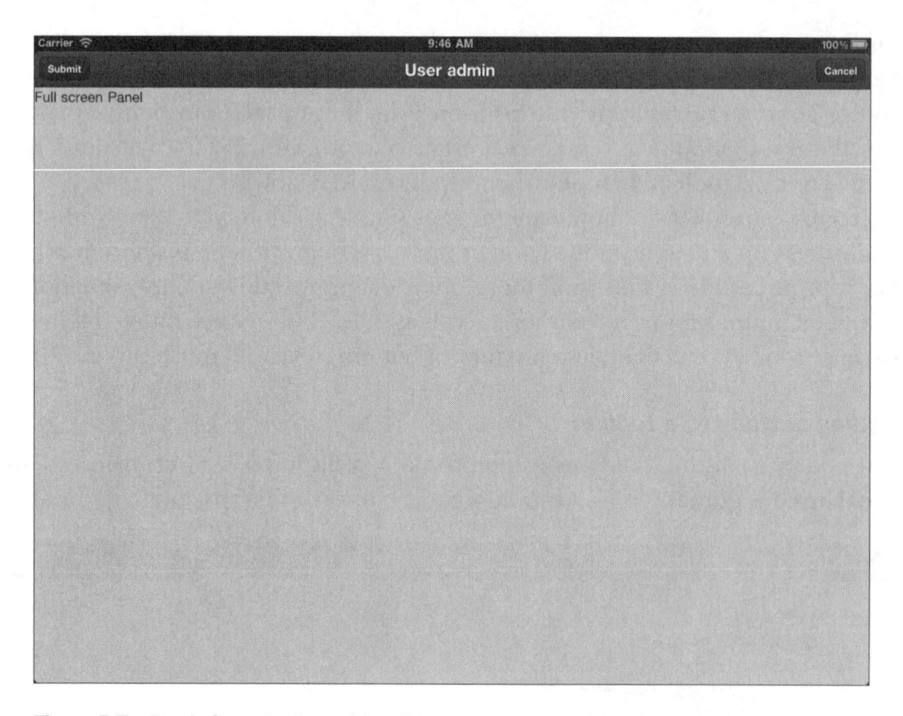

Figure 5.7 Rendering a toolbar with a title, two buttons, and a hidden spacer component

This is because the spacer component takes 100 percent of the available screen space between the two buttons.

To get a full grasp on what's happening here, recall chapter 4, where you learned about the `HBox` layout and the `flex` configuration property. By default, the toolbar uses the `HBox` layout internally, whenever it's docked at the top or bottom. The reason the buttons are pushed to opposite ends is because the spacer component is between them, and it has a default `flex` property of 1, instructing the `HBox` layout to size the spacer to 100 percent of the available horizontal space minus the widths of the buttons and some padding.

5.2.3 *Centering items*

The fact that the toolbar implements an `HBox` layout can be extremely useful if you want to center buttons or a set of buttons. The following listing modifies listing 5.4 and is a recipe for centering an instance of `SegmentedButton` between our two previous buttons.

Listing 5.5 Centering with two spacer components

```
var toolbar = {
    xtype  : 'toolbar',
    docked : 'top',
    items  : [
        {
            xtype : 'button',
            text  : 'Submit'
        },
        {                                        ❶ Contains first spacer
            xtype : 'spacer'
        },
        {
            xtype : 'segmentedbutton',           ❷ Configures SegementedButton
            items : [
                {
                    text : 'Comedy'
                },
                {
                    text : 'Thrillers'
                },
                {
                    text : 'Horror'
                }
            ]
        },
        {
            xtype : 'spacer'                     ❸ Contains second spacer
        },
        {
            xtype : 'button',
            text  : 'Cancel'
        }
```

```
    ]
};

Ext.create('Ext.Panel', {
    fullscreen  : true,
    style       : 'background-color: #CCF;',
    html        : 'Full screen Panel',
    items       : toolbar
});
```

Listing 5.5 demonstrates how to center a `SegementedButton` ❷ by sandwiching it between two spacer (❶ and ❸) components. Figure 5.8 shows how it looks rendered onscreen.

The result of listing 5.5 is shown in figure 5.8, where a `SegmentedButton` is centered between two outside buttons. The reason it's centered is because the spacer has a default `flex` property of 1, instructing the `HBox` layout to allow the spacers to use all the available space equally.

Because of the `HBox` layout's flexibility you can set static widths on the spacers or even have the buttons dynamically sized via the `flex` parameter. The `HBox` layout's flexibility can lead to truly creative toolbar item organization, but as we said earlier, you can render just about anything inside the toolbar.

What you just implemented was the centering of items in addition to having items flush on each side of the toolbar. If you wanted to have all items centered you wouldn't need to use the flexible spacer component but instead would simply instruct

Figure 5.8 Centering a `SegmentedButton` by means of two spacer components

the `HBox` layout to do the work for you by adding the following parameter to your `Toolbar` configuration object:

```
layout : {
  pack : 'center'
}
```

You've just experimented with various ways to lay out items inside a toolbar. Next, you'll take a stab at a toolbar configuration that can implement more than just buttons and spacers.

5.2.4 Adding nonstandard components

Because you've already learned that `Toolbar` extends `Container` and thus allows for any item to be placed in it you're going to proceed with a recipe for a toolbar-based search for your application. To do this you'll need to add a text input field to the toolbar `items` configuration parameter, as shown in the following listing.

Listing 5.6 Adding a text input field to a toolbar

```
var toolbar = {
    xtype  : 'toolbar',
    docked : 'top',
    title  : 'User admin',
    items  : [
        {
            xtype : 'spacer'                   ❶ Contains flexible spacer
        },
        {
            xtype : 'textfield',               ❷ Configures textfield
            width : 200
        },
        {
            xtype    : 'button',               ❸ Adds search button
            iconCls  : 'search',
            ui       : 'plain',
            iconMask : true
        }
    ]
};

Ext.create('Ext.Panel', {
    fullscreen  : true,
    style       : 'background-color: #CCF;',
    html        : 'Full screen Panel',
    items       : toolbar
});
```

Listing 5.6 proves that you can render more than just buttons in a toolbar. Because a common design paradigm is to place search fields to the right of the toolbar, use a spacer component to move things over ❶. Next is the text ❷ input field, followed by a button ❸ configuration option that has some unique styling. Figure 5.9 illustrates how it renders onscreen.

Figure 5.9 The text input field rendered inside a toolbar, alongside a nicely styled button

So far you've seen how toolbars work, but we haven't talked much about buttons. It's hard to argue with the fact that buttons are an integral part of many of the screens you'll be creating for any application. This is why we'll be shifting gears to focus on these versatile widgets.

5.3 *Go ahead, press my button!*

The Sencha Touch `Button` class is used all over the place in your applications, and for whatever reason, its versatility is often overlooked by many of us. The fact of the matter is that the button can be easily customized to match almost all of your application needs.

The general use of buttons is simple. For example, here's a typical `Button` configuration object:

```
var myBtn = {
    xtype     : 'button',
    text      : 'Delete',
    iconCls   : 'delete',
    iconMask  : true,
    ui        : 'drastic',
    scope     : someObject,
    handler   : function() { /* code here */ }
}
```

This configuration object can be used to render a Button within a Toolbar or Container, though the xtype property is typically omitted when being used inside a toolbar, because its default type is button.

When tapped the button will execute a function, known as a handler, which can be executed within the scope of any object.

> **Does "scope" sound foreign?**
> In JavaScript functions can be executed within the "scope" or context of any object. This means that when the function is executed the magic keyword this points to said object. To demystify the this keyword visit www.slideshare.net/moduscreate/javascript-classes-and-scoping.

At runtime you can change the handler of any button via its setHandler method, which takes two arguments, a reference to a function that will be called on the button's tap event and the scope in which that function will execute.

5.3.1 Customizing buttons

Buttons can be rendered anywhere on screen, and they can be children of any Container or subclass of Container. The next listing shows an example of some modified buttons rendered in a top-docked toolbar, as well as inside a panel. Because of the sheer number of configuration properties in this example it's going to be pretty long, so please bear with us.

Listing 5.7 Customizing buttons

```
var tbar = {                                  // 1 Configures toolbar
    xtype    : 'toolbar',
    docked   : 'top',
    defaults : {
        iconMask  : true
    },
    items : [                                 // 2 Adds buttons
        {
            text      : 'delete',
            iconCls   : 'delete',
            iconAlign : 'left',
            ui        : 'back'
        },
        {
            text      : 'organize',
            iconCls   : 'organize',
            iconAlign : 'right',
            ui        : 'decline'
        }
    ]
};
```

```
var bbar = {
    xtype    : 'toolbar',
    docked   : 'bottom',
    defaults : {
        iconMask : true
    },
    items : [
        {
            iconCls   : 'refresh',
            iconAlign : 'top',
            ui        : 'plain'
        },
        {
            text      : 'search',
            iconCls   : 'search',
            ui        : 'drastic',
            iconAlign : 'bottom'
        }
    ]
};

Ext.create('Ext.Panel', {
    fullscreen  : true,
    defaultType : 'button',
    defaults    : {
         iconMask : true
    },
    layout : {
        type : 'vbox',
        pack : 'center',
        align: 'center'
    },
    items : [
        tbar,
        bbar,
        {
            text      : 'generic',
            width     : 150
        },{
            text      : 'compose',
            iconCls   : 'compose',
            iconAlign : 'right',
            ui        : 'action',
            width     : 150
        },{
            text     : 'star',
            iconCls  : 'star',
            ui       : 'confirm',
            width    : 150
        }
    ]
});
```

❸ Docks toolbar to bottom

❹ Adds buttons to bottom toolbar

❺ Includes both toolbars

❻ Adds buttons to panel

Listing 5.7 contains the necessary code to create a panel that contains both bottom- and top-docked toolbars as well as three buttons rendered in its body. Here's how it all fits together.

You begin by creating a configuration object for the top-docked toolbar ❶. This toolbar contains two buttons ❷, with an icon alignment set via the `iconAlign` property.

> **Know thy iconAlign**
> The `iconAlign` property can be set to `left`, `right`, `top`, or `bottom`.

Because you're using the built-in icons, both of the buttons must have their `iconMask` option set to `true`. Setting this option to `true` instructs Sencha Touch to apply the default `iconMaskCls` CSS class (`x-icon-mask`) to the button, applying a CSS alpha channel mask that makes the icon visible on the screen. If it's not set to `true` the icons won't be visible.

> **Learn about CSS alpha channel masks**
> CSS alpha channel masks allow you to take one image (the alpha channel) and blend it with another to create complex blending that previously required powerful image editing software. For an excellent beginner's tutorial on masks visit www.webkit.org/blog/181/css-masks/.

Next, configure the bottom-docked toolbar ❸ that also has two buttons ❹. One of the buttons has its icon top-aligned, whereas the other button has its icon bottom-aligned. As you'll see in a bit, the icons will render above and below the text according to the `iconAlign` configuration property. Notice that the `iconMask` property is set to `true` via the toolbar's `defaults` configuration object.

> **Unleashing the pictos (icons)**
> A little-known fact is that Sencha Touch has over 330 icons that can be used in your mobile apps, but only 26 of them are usable out of the box. To access the rest you'll need Sass and the related libraries, such as Ruby, and some elbow grease. To download Sass you can visit http://sass-lang.com/. Also Shea Frederick, a veteran of the Sencha community, has an excellent article on his blog that details how to use Sass to compile the Sencha Touch CSS to include only the icons you plan on using in a process he calls "trimming the fat." His article can be found at www.vinylfox.com/custom-styling-a-sencha-touch-app/.

The last block of code in listing 5.7 creates a full-screen panel ❺ that contains three buttons ❻ centered on screen by means of the `VBox` layout.

There are some other bits of information about this listing that we should discuss, but we think it's best to see the result first, because that will better frame up what we'll be talking about next. Figure 5.10 shows how it all looks rendered on a mobile device.

Figure 5.10 Rendering customized buttons as a result of listing 5.7

Figure 5.10 illustrates the result of listing 5.7, rendering a panel containing a top-docked and a bottom-docked toolbar with two customized buttons each. Notice how the icon alignment is organized exactly how you configured it to be.

Each of the buttons has something unique aside from `iconAlign`-ment, and that has to do with the `ui` configuration property. Of the five buttons rendered onscreen, one of them is a generic button, meaning it doesn't have an icon and doesn't have a `ui` configuration property to alter its appearance.

This optional configuration property is important to remember because it allows you to alter the appearance of buttons and other widgets with ease. Here's how it works.

When you specify the `ui` configuration property Sencha Touch adds a CSS class to the widget's element, allowing you to alter its physical appearance by using some of the built-in UI styles. The two buttons in the top-docked toolbar in figure 5.10 have `ui` properties set. The first of these properties is `back`, which alters the physical shape of the button but doesn't alter the default color. The second button uses the `decline` `ui` style, which gives it a red color but doesn't change its shape.

If you look at the rendered buttons in the panel you'll see that there are three colors. The first is a generic button and demonstrates how buttons will appear if rendered outside of a toolbar without icons or a `ui` style applied. The second uses a `ui` style called `action`, giving it a blue tint, whereas the third button uses the `confirm` `ui` style, coloring it green.

Focusing on the bottom toolbar, you'll find a button that has no border around it and a generic button with `iconAlign` set to `bottom`. The focal point here is the first button, which has its `ui` property set to `plain`. Setting a button's `ui` property to `plain` within a toolbar will render the button without borders. We must stress the fact that the `plain` `ui` style only works within toolbars. Adding this `ui` style to a button rendered inside a container will prevent the button from rendering.

In addition to the `back` `ui` style you can alter the shape of buttons via the `forward`, `round`, and `small` `ui` styles. You even have the capability to use compound styles, such as `action-small`, `confirm-small`, and `decline-small`.

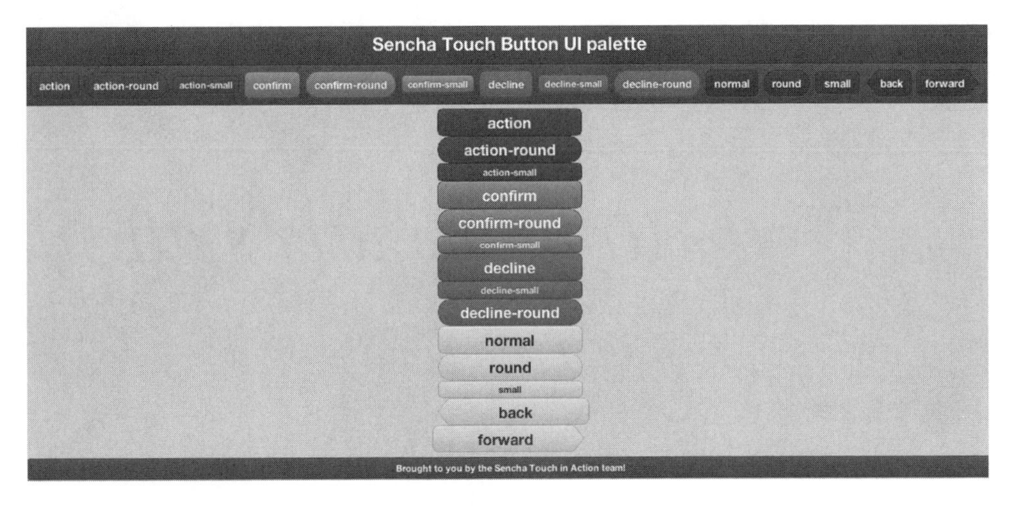

Figure 5.11 Demonstrating all possible button styles

Instead of listing all of the possible combinations, figure 5.11 shows a quick example of all the possible out-of-the-box styles.

Figure 5.11 shows all of the possible out-of-the-box button `ui` styles within a toolbar and a panel. The `ui` style used to alter each button is set as its text, so you can quickly figure out what style you want to use. In case you're wondering, the code for this example comes with the example code zip file under examples/chapter04/all_button_styles.html.

Now you know just about everything you need to know about how buttons work and how to deploy them in a normal or customized manner.

5.4 Summary

Great work! We covered a lot in this chapter and you worked on many of the building blocks for mobile application development with Sencha Touch.

You began by looking into some of the intricacies behind docked items and where they can be placed, and learned the importance of the order of docked items and its effect on item sizing. From there, you learned about the most commonly docked item, the toolbar, and saw how to arrange and space out items within toolbars. Last but not least, you explored how to customize buttons and make them a bit more user-friendly through various icons and styling options. All of this work set the stage for you to explore how Sencha Touch uses many of these concepts in the built-in widgets it ships with. The next chapter focuses on sheets, pickers, and the message box, all of which use various concepts from this chapter and the previous ones.

Getting the user's attention

6

This chapter covers

- Creating simple overlays
- Using sheets as modal dialogs
- Understanding the `Picker` class
- Using and customizing message boxes

While exploring the exciting world of containers, panels, and multiscreen applications via the use of components and tabs in the previous chapter, you might've noticed that all of those paradigms share one trait: they all force the user away from the current application screen to facilitate user interaction. Although this is frequently a desired behavior, especially if content needs to be shown in a logically separated way, navigating between different screens can disrupt the flow and usability of your app. This is particularly true if all you need is to get a quick message to the user, or prompt for simple information like a name. In such cases a TabPanel or a card layout that navigates to a different screen simply to return to the user after information has been gathered is just overkill.

This is exactly where sheets, pickers, and message boxes come into play. These components are specifically designed to allow quick interaction with the user and enable you to easily overlay information onto the current screen (sheet), allow the user to select something from a list of values (picker), or notify the user of something that has happened (message box). Each component serves its purpose, and understanding the ins and outs of all three components and how they fit together is necessary. Let's start our journey through these components by taking a look at sheets.

6.1 Using sheets for modal user interactions

Sheets form the basis for all modal dialogs in Sencha Touch. Whether you're looking at a date picker, a select field, a message box, or an ActionSheet, they all extend the basic `Sheet` class. To see exactly how all the modal dialogs are based on the `Sheet` class take a look at the class diagram in figure 6.1.

> **For those developers coming from Ext JS**
> You can think of `Sheets` as a close cousin to the `Ext.Window` class with a modal option set to `true`. The main difference is that sheets are panels and have all the same options as standard panels. These include layouts, toolbars, and child item management.

By default sheets slide in from the bottom of the screen, automatically covering up existing content and imposing a mask on the rest of the screen. Although this is the default, you aren't limited to this behavior only. With a few simple tweaks to the configuration the sheet can be made to enter from any direction using the `enter` property. It

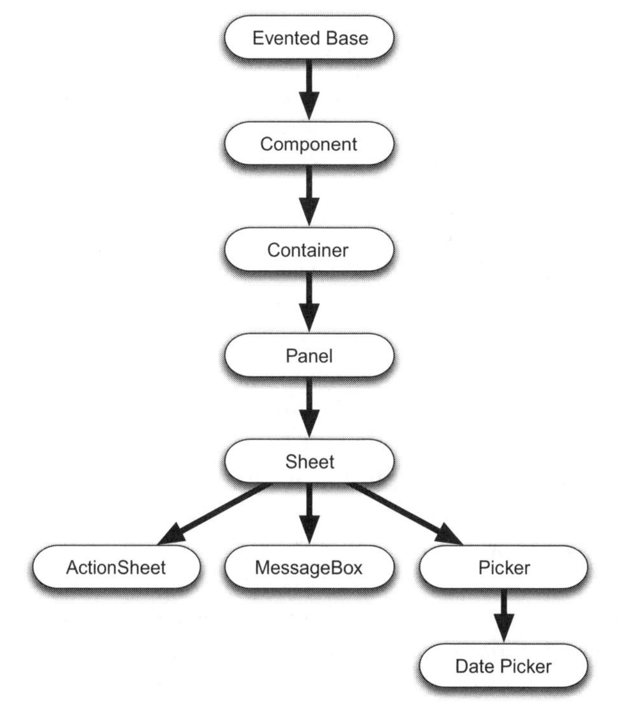

Figure 6.1 Sheets and modal dialogs class diagram

Figure 6.2 A sheet sliding in from the bottom on a phone (left) and the same sheet rendered on a tablet (right)

could be made to slide in from the left side of the screen, for example. The beauty behind sheets is that no matter which behavior you choose the sheet will behave the same way on all devices, causing it to work seamlessly on phones and tablets alike. Take a look at figure 6.2 to see a sample of a fully rendered `Sheet` instance on a phone and a tablet.

As you can see, the same sheet rendered on two different devices shows the same content in the same place, only utilizing more space on a tablet than on a phone. Reorienting a device keeps the sheet in its place and automatically recalculates its size.

Size is where one of the weaknesses of sheets comes to light. Unless you specify a `height` and `width` parameter or utilize the stretching configuration options (`stretchX` and `stretchY`, respectively), your sheets won't automatically size themselves properly. Because sheets are panels they use the default `AutoContainerLayout` discussed in chapter 4, rendering components one on top of the other without any care for sizing or restrictions on the height of the sheet. Leaving off the `height` can result in a sheet that looks like figure 6.3, where the content isn't shown and all that's visible is the toolbar.

On the flip side, adding too many elements without specifying a proper layout can result in one of the two scenarios shown in figure 6.4, where either the content extends beyond the sheet or part of the sheet reaches beyond the screen, inaccessible to the user.

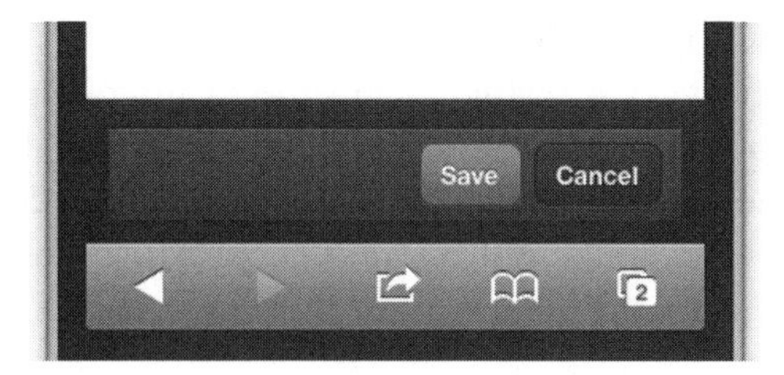

Figure 6.3 A sheet without proper `height`

The moral of the story is that you should always make sure you specify the proper `height`, `width`, and `layout` configuration to avoid situations like this.

That's enough theoretical talk for the moment. It's time you get your hands dirty and start building some sheets.

6.1.1 Using sheets for simple overlays

In its most basic form a sheet can be used as a straightforward overlay, triggered to show additional information about content on the screen.

> **For those developers coming from Ext JS**
>
> You can think of sheets as a less powerful version of the `Ext.Tooltip` class that exists on the Ext JS side. On the Ext JS side things like mouseovers are possible, and close buttons are built in from the start in an aesthetically pleasing way. In comparison, sheets in Sencha Touch are bound to finger input and require you to add your own close button and style it.

Figure 6.4 Sheets without proper layout configuration

In general it'd probably be advisable to use a floating panel to represent the overlay because that provides functionality for `show by`, which aligns the floating panel with a specified component or element. In certain instances design might dictate that overlays are shown by sliding in from the bottom of the screen (like sheets do) for space consideration or any other arbitrary reason. The following listing illustrates how to do this.

Listing 6.1 Using a sheet as an overlay

```
var buttonHandler = function() {
    Ext.Viewport.add({
        xtype          : 'sheet',
        height         : 130,
        hideOnMaskTap  : true,
        hidden         : true,
        enter          : 'bottom',
        style          : 'color:white; ',
        html           : 'Place your additional information right here.'
                         + '<br />It can be multi line and everything!',
        items          : {
            docked   : 'bottom',
            xtype    : 'button',
            text     : 'Close',
            handler  : function(button) {
                button.up('sheet').hide();
            }
        }
    }).show();
};
Ext.create('Ext.Panel', {
    fullscreen : true,
    layout     : {
        type   : 'vbox',
        align  : 'center',
        pack   : 'middle'
    },
    items      : {
        xtype   : 'button',
        text    : 'Open the overlay',
        handler : buttonHandler
    }
});
```

❶ Adds sheet to viewport, shows it

❷ Adds button to close sheet

❸ Creates panel with button

❹ Handles click events on button

Probably expected more, eh? Fortunately, the code ends up on the simple side. To understand how this works you need to first focus on the bottom portion of the listing, where you render a full-screen panel ❸ with a centered button ❹. This button has a handler configured, known as `buttonHandler` ❶ in this program.

`buttonHandler` is responsible for creating an instance of `Sheet` by using the viewport's `add` method and passing a lazy instantiation object to it. The sheet is shown because you chain a `show` method call immediately after the add operation. The sheet has an added close button ❷, which has an inline handler to hide the sheet via an

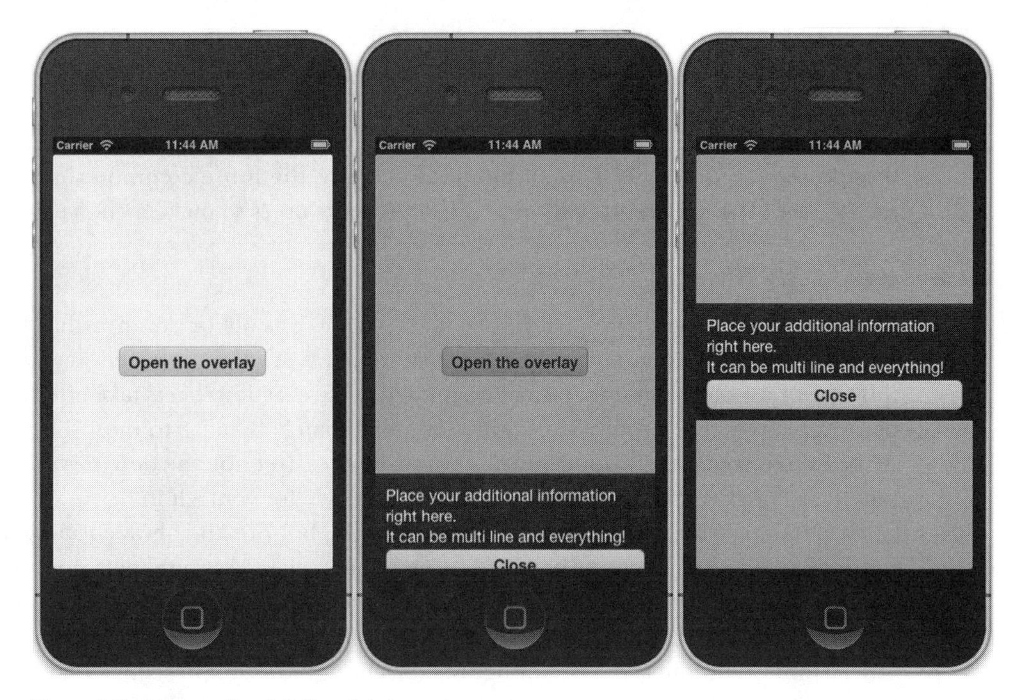

Figure 6.5 The results of listing 6.1

upward component query (using the `Button` instance) to find the parent `Sheet` class and chain a `hide` method call. Figure 6.5 shows what it looks like rendered on a phone, animating in the `Sheet` instance that you create via the `buttonHandler` function.

The sheet you created earlier is set up to slide in from the top, which is accomplished with the `enter` property. Changing this property to `left`, `right`, or `bottom` changes where the sheet enters the screen. To ensure the sheet occupies the full width of the screen you set it to stretch on the x-axis via `stretchX: true`. For content you populate it with some text through the `html` property. By default the sheet doesn't understand when it needs to hide. This means you either have to add a toolbar with a close button on it or set it to automatically hide whenever the mask (the area not occupied by the sheet) is clicked via the `hideOnMaskTap` property. Notice that in order to show the sheet you need to add it to the overall viewport. You must do so in order that the sheet can be managed for layout purposes. One thing to keep in mind here is that once you add the sheet to the viewport it'd normally show up automatically (without any animation). To fix this you set the sheet to `hidden: true` in order to give you full control over when it should be shown.

All in all, the design pattern is deceptively simple. Where things get interesting is dealing with content that needs to be dynamically loaded into a sheet. In that case, the elements in your panel body would each need a unique identifier so you can determine which specific element was clicked. From the event handler you'd make your Ajax call to the backend to retrieve whatever data you need, waiting for the success handler to come back before showing the content of the sheet. This way, no sheet would show if the Ajax call failed, and you can fish the content out of the response and dynamically set it before creating the sheet. Voilà!

Although using sheets as overlays is a nice example it most likely represents one of those cases you don't run into all too often. One of the more common sheet types is the ActionSheet, which mimics the way iOS prompts users to make a choice.

6.1.2 Using ActionSheets

Whether you need to prompt the user for a yes/no answer or confirm that the user wants to delete that one record in a list, ActionSheets are the preferred way to handle this type of user interaction. From a design perspective ActionSheets take after the standard Apple iOS style prompts and will look immediately familiar to most iPhone/iPad users. Even if you're developing for non-Apple devices, fret not; the design is simple and intuitive enough to be self-explanatory to the user. See for yourself in figure 6.6.

By default, ActionSheets always slide in from the bottom and always occupy the full width of the screen, and just like the normal `Sheet` class, ActionSheets don't change their looks or behaviors based on the user's device. This means that the ActionSheet from the left side of figure 6.6 will look exactly the same on a tablet except that it'll be wide, as shown on the right side of figure 6.5. Though this might seem a bit counterintuitive at first, especially given the aesthetics of the tablet version, it does create a sense of consistency across different devices and removes any doubt as to the meaning of the ActionSheet.

Setting up an ActionSheet is relatively straightforward, and the next listing illustrates how you'd create the ActionSheet in figure 6.6.

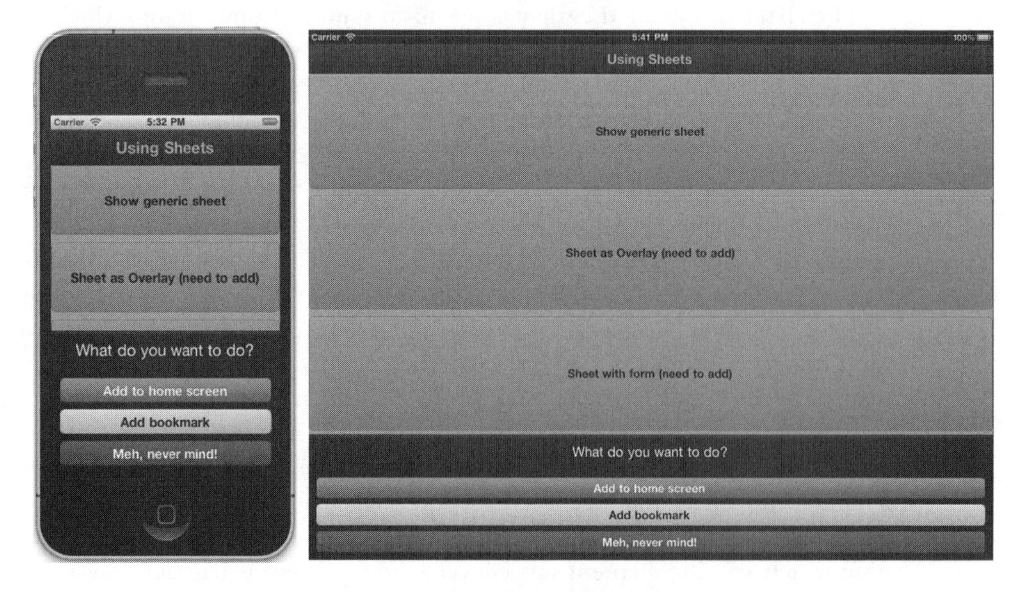

Figure 6.6 A simple ActionSheet with three choices and a title

Listing 6.2 A simple ActionSheet

```
var myActions = Ext.create('Ext.ActionSheet', {
    items : [
        {
            xtype  : 'component',
            height : 50,
            style  : 'color: #FFF; text-align: center; font-size: 1.2em;',
            html   : 'What do you want to do?'
        },
        {
            text : 'Add to home screen',
            ui   : 'confirm'
        },
        {
            text : 'Add bookmark'
        },
        {
            text : 'Meh, never mind!',
            ui   : 'decline'
        }
    ],
    defaults  : {
        handler : function() {
            this.ownerCt.hide();
        }
    }
});
Ext.Viewport.add(myActions).show();
```

❶ Uses html for title
❷ Defines actions
❸ Sets up default handler

You begin by setting up an `Ext.ActionSheet` instance with a title and three items. In this implementation you're opting to put the title in a standard component ❶ using `html` and placing it in the `items` array of the ActionSheet instead of a docked title bar.

The visual difference might not be immediately apparent until you compare the title bar from figure 6.5 with that of figure 6.4. A docked title bar creates a starker visual separation between the title and the content, primarily due to the different background and spacing between the toolbar and the content. By using a simple component you're sacrificing the functionality that a toolbar would provide you in exchange for a title that looks more like it's part of the content.

Next you add the three actions ❷ into the `items` array. Because the default item type for `ActionSheet` is `Button` and all actions are buttons, you don't need to specify an `XType` for your actions. Each action receives a title via the `text` property and an optional `ui` parameter to visually distinguish it. Notice that none of the buttons have a handler defined. In this state they'd simply be dummy buttons that don't do anything. You take care of that through the use of the `defaults` object and provide a default handler ❸ for all actions that hides the ActionSheet. The handler isn't scoped in any specific way so the `handler` function will receive the default scope: the button that was clicked. You don't want to hide the button so you use the convenient `parent` property, which gives you access to the button's parent, the ActionSheet. Keep in mind that this setup is simply for illustrative purposes. In a real-world application, each item would most likely have a unique handler that performs a particular action instead of all actions sharing the same handler.

One caveat to keep in mind when dealing with ActionSheets is that you can define an arbitrary amount of actions and no limit is enforced. Although this is great in cases where you need to present a handful of options, it can quickly get out of hand if your actions exceed the available screen space because the ActionSheet doesn't support scrolling by default. In cases where you need to present a particularly large set of items to the user simply use a picker, which we'll cover in the next section.

6.2 Choosing pickers

Imagine sitting at Caesar's Palace in Las Vegas, drink in one hand and the other feeding a one-armed bandit. You pull down its arm and the slots start spinning. The first slot shows a 7; then a second 7 shows. One more and you hit the jackpot. The third slot is spinning for what seems like an eternity. It stops, and...you lose yet again.

Pickers in Sencha Touch have a similar concept to slot machines; they provide a series of "slots" with a predefined set of values for each slot. But unlike in Vegas, Sencha Touch allows users to pick the exact value they want. For a sample of this take a look at the screen in figure 6.7.

The picker behaves just like all other classes in the `Sheet` family, where the default behavior is to slide in from the bottom of the screen when triggered and occupy the full width of the screen, behaving the same way on phones and tablets alike. Internally, Sencha Touch uses a picker for the select form field, which we'll cover in more detail in chapter 7. But let's not get ahead of ourselves; first let's explore basic pickers like the one in figure 6.7.

Figure 6.7 A typical picker with a single "slot"

6.2.1 Creating a basic picker

Under the hood the picker consists of a sheet with multiple `Ext.Picker.Slot` instances, where each slot is a scrollable data view. A picker can have an arbitrary number of slots, limited only by the amount of screen real estate available to you. The following listing demonstrates how to go about building a basic picker.

Listing 6.3 A basic picker

```
var myPicker = Ext.create('Ext.Picker', {
    useTitles : true,
    hidden    : true,
```

1 Generates titles

```
    slots      : [
        {                                                    ❷ Sets up slot
            name  : 'agegroup',
            title : 'Age group (years old)',
            data  : [                                        ❸ Provides data for slot
                { text: '1 - 10', value:  1 },
                { text: '11 - 14', value: 2 },
                { text: '15 - 18', value: 3 },
                { text: '19 - 24', value: 4 },
                { text: '25 - 31', value: 5 },
                { text: '32 - 40', value: 6 },
                { text: '41 - 50', value: 7 },
                { text: '51 - 75', value: 8 },
                { text: '75 - 100+', value: 9 }
            ]
        }
    ]
});
Ext.Viewport.add(myPicker).show();
```

You start out with the simplest picker possible, containing only a single slot. You instantiate the picker and tell it to give each column a title ❶ via the `useTitles` property. When this property is enabled the picker will automatically add a toolbar to each slot with a title equal to the `title` property defined in the slot. In this case the title would be “Age group (years old)”. You provide the one and only slot ❷ defined for this particular picker. The picker automatically provides the `XType` for the slot, so you don’t have to worry about it at all. Giving the slot a name provides you with a way to identify each slot and its corresponding value if you were to call the `getValue` method of the picker. The only thing left is to define the data to show. You do so via an array of objects ❸, each of which consists of a `text` property and a `value` property. The former is the actual text that shows in the picker and the latter is the value that’d be returned as part of the `getValue` call.

If you wanted to add another slot to the picker you’d build it adjacent to the current one and structure it the same way, with a name, a title, and an array of data objects. That’s all there really is to pickers.

When looking at the documentation for pickers in the Sencha Touch docs you’ll most likely notice a severe deficiency in styling options. The standard picker looks somewhat bland. One bright spot is that you can easily add HTML tags to the text. Tags allow you to provide some quick-and-dirty, albeit not dynamic, styling for each item. Here’s a sample `slot` definition:

```
{
    name : 'color',
    title: 'Color',
    data : [
        { text: '<div class="color-red">RED</div>', value: 'red' },
        { text: '<div class="color-green">GREEN</div>', value: 'green' },
        { text: '<div class="color-blue">BLUE</div>', value: 'blue' }
    ]
}
```

This snippet assumes that the following CSS classes are present:

```
.color-red { background: red; }
.color-green { background: green; }
.color-blue { background: blue; }
```

In this case each item in the slot would show with a different color: red, green, and blue respectively, just like in figure 6.8.

Figure 6.8 A picker with styled HTML content

Although you might cry foul, claiming that this approach is cheating and doesn't style the picker itself, the truth of the matter is that pickers don't provide much out-of-the-box functionality to change their appearance. Sadly, most of the functionality gained from the fact that slots extend data views has been locked down, preventing you from passing a custom XTemplate, for example. To truly change the look and feel of a picker you'll have to venture deep into the bowels of Sencha Touch and write your own extension to the `Picker` and `Slot` classes, as well as some serious CSS.

On the topic of extending classes, let's see if you can get a better idea of what's achievable by taking a look at the date picker to extend the standard `Picker` class.

6.2.2 Date picker

Built on top of the standard `Picker` class, the built-in date picker consists of three slots, one each for year, month, and day. Onscreen the entire thing looks pretty much the way you'd expect a picker with three slots to look (figure 6.9).

Things get interesting when you take a closer look at the configuration options and the actual code behind the date picker. To better illustrate the options available to you take a look at the code for figure 6.9:

```
var myDatePicker = Ext.create('Ext.DatePicker', {
     yearFrom  : 2010,
     yearTo    : 2025,
     slotOrder : ['year', 'month', 'day'],
     value     : {
         year   : 2011,
         month  : 7,
         day    : 4
```

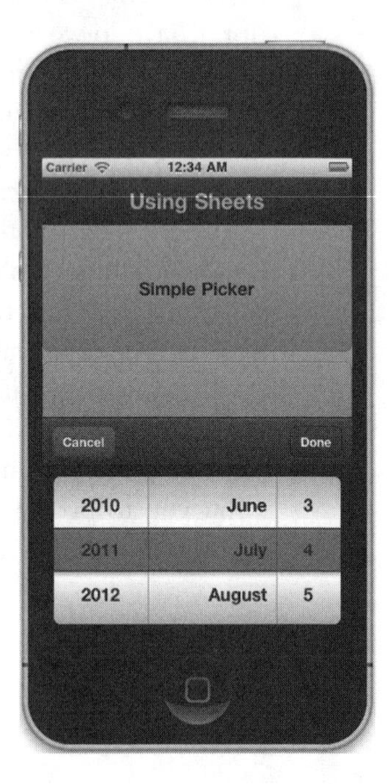

Figure 6.9 A standard date picker

```
    }
});
Ext.Viewport.add(myDatePicker).show();
```

In this case you limit the dates to start in 2010 (via the `yearFrom` property) and only go up to 2025 (via the `yearTo` property). By default the year starts in 1980 and goes until the current year. Slots are normally sorted by month/day/year, but because you want to be friendly to your international readers you'll change the order. To do so you utilize the `slotOrder` property and change it to year/month/day instead. The date picker takes care of automatically creating all the slots for you, so you don't have to worry about that part. The only thing left is to set a starting value. This can either be a proper JavaScript date object or a JSON object with a `year`, `month`, and `day` property.

If you encounter one of those cases where you only need to query a user for a year and a month, for example, you can simply change the slot order and leave out the day slot. This will automatically hide any slot not mentioned in the order.

Calendar picker is MIA

For those developers coming from the Ext JS side, as of this writing the date picker is the only way to select dates and there's no true equivalent to the Ext JS date picker that shows a calendar-like structure split into rows and columns.

Although pickers are easy to set up and sheets are versatile in the content you can populate them with, neither one provides an easy or quick way to prompt the user for input or relay a quick message to the user. Fortunately, the Sencha Touch developers have thought exactly of this use case and given you the `MessageBox` class.

6.3 Talking to the user via a message box

Whether your goal is to notify the user of something that happened, prompt for input, or query the user to make a quick yes/no decision, the `MessageBox` class is the way to go. It provides the easiest out-of-the-box interaction with the user.

`MessageBox` extends the basic `Sheet` class and automatically sets a few of the configuration options associated with the `Sheet` for you. `MessageBoxes` by default always use the `dark ui`, show `centered` on the screen, and pop in and out for their `enterAnimation` and `exitAnimation`.

Creating a message box largely follows the same paradigm as all the other Sencha Touch components, but there are a few caveats to keep in mind that distinguish it from the rest.

First, you as a developer never have to worry about instantiating a `MessageBox` instance yourself. Although it's possible to do so Sencha Touch automatically creates a singleton instance for you with the global alias `Ext.Msg`, thus removing the need for you to create your own before using it. `MessageBox` is the only component in the framework that does this.

What this means effectively is that instead of doing this

```
var myMessageBox = Ext.create("Ext.MessageBox", { /*options here*/ });
myMessageBox.show();
```

you can easily do this:

```
Ext.Msg.show({ /*options here*/ });
```

The primary reason for this is that the heart of `MessageBox` is in its `show` method, which allows for a rather complex configuration object to be passed to it. Before we delve into the depths of the `show` method let's take a look at the simplest way to show a message box via the following sample:

```
Ext.Msg.show({
   title: 'Southpark',
   msg: 'Ice Ice Timmy!'
});
```

This code uses the `MessageBox` singleton to show a simple message consisting of a `title` and `msg` (message) body. If you run the code from the console you'll notice the presence of an OK button, just like you'll see in figure 6.10.

Even though you didn't specify it the message box automatically creates a toolbar with an OK button to provide the user with an easy way to dismiss your message. Given that this example barely scratches the surface of what's possible let's see if you can get a better idea of your options from the `show` method and some of the helper functions, starting with alerts.

Figure 6.10 Basic `MessageBox` with title and message body

6.3.1 Alerting users

Alerts consist of a title and a message body, and they're perfect for telling the user when an operation has finished or an error has occurred, or any similar type of message. The outcome is much like you've already seen in figure 6.10.

There are only three properties within the `show` method of `MessageBox` that concern themselves with showing text. Table 6.1 provides details on all of them.

Table 6.1 The available `show` options for displaying basic text

Option	Description
`title`	This option sets the title of the message box. If it's left blank or omitted entirely no top toolbar is created. As of this writing it's impossible to have a multiline title, either by entering long text or by forcing a line break via HTML tags.

Table 6.1 The available show options for displaying basic text *(continued)*

Option	Description
msg	This option specifies the message body you want to show. In the event of a prompt you'll use this as the text to show right above the prompt, similar to a label for a text field.
modal	This Boolean option defines whether the message box should be modal. It determines whether the rest of the screen should be masked or still be available for interaction.

The code that creates an alert is the same as the code you used for figure 6.9 earlier. For convenience purposes the `MessageBox` class exposes an `alert` function, which automatically wraps the `show` method with default values. To use the `alert` method follow this sample:

```
Ext.Msg.alert("Title Goes Here", "Message goes here");
Ext.Msg.alert("Title Goes Here", "Message goes here", myAwesomeFn, this);
```

The `alert` function is broken down into four parameters: the first is the title, the second the message to show, and the third and fourth a callback function to invoke when the message box is closed and its respective scope. Under the hood the `alert` method automatically inserts an OK button and calls the `show` method, passing along the specified parameters. In the previous code the first function call would simply show the alert with the provided `title` and `msg`. In the second call the `myAwesomeFn` function would be invoked when the OK button is clicked, using the scope of whichever object called the alert. The callback function passes you three parameters. The first is the `itemId` of the button that was clicked. In this case it would be `OK`. The second is the value, which only applies to prompts, and in the case of an alert is null. The last is a copy of the config object passed to the `show` method. For the most part, the callback function is used more infrequently on alerts than any of the other types of message boxes. To see a better example of the callback being used let's take a look at the next section, which deals with ways to prompt the user for simple yes/no answers and then act based on those answers.

6.3.2 *Prompting users*

In the world of Sencha Touch a message box that presents the user with a set of buttons to be clicked is called a confirmation box. Typically confirmation boxes are used in cases where the user might do something risky, and you as the developer have to protect the user from rash actions like deleting a record, submitting a payment form, or other similar things.

In terms of coding confirmations follow the same paradigm as the options already used for alerts in figure 6.9, with the addition of configuration options to define the buttons and callback function. The easiest way to invoke a confirmation message is via the `Ext.Msg.confirm` method. The syntax is exactly the same as for the `alert` method discussed earlier:

```
Ext.Msg.confirm(
    "Are you sure?",
    "This will shutoff the internet!",
    myAwesomeFn
);
```

Figure 6.11 Default confirmation box

The primary difference between `alert` and `confirm` is a change in buttons. Where the alert method only shows an OK button the confirmation box by default shows a Yes and a No button, just like you'll see in figure 6.11.

For the confirmation box to be truly useful you'll have to make sure you pass a callback function so you can process which button was actually clicked. Because the `confirm` call depicted in figure 6.11 expects a callback named `myAwesomeFn` let's see what that might look like:

```
var myAwesomeFn  = function (buttonID, value,
     opts) {
     if (buttonID === "no") {
          // whew. Threat averted!
     }
     else {
          // oh noes! the internet is going down
     }
}
```

This is all that's needed to handle a confirmation box and process the response.

"Wait!" you might say. "This can't be everything!" You know for a fact you've seen message boxes with multiple buttons that weren't Yes or No. Right you are. The Yes/No combination is simply the default when using the `Ext.Msg.confirm` helper function. To define your own buttons you'd use the `show` method and pass in your own buttons configuration, just like this:

```
Ext.Msg.show({
    title   : 'Are you sure?',
    msg     : 'This will shutoff the internet!',
    buttons : Ext.MessageBox.OKCANCEL,
    fn      : callbackFunction
});
```

This code would show the same confirmation prompt as figure 6.11, but the buttons would be OK/Cancel instead of Yes/No. To better understand the various options related to confirmation dialogs and how to define custom buttons take a look at table 6.2.

One often-overlooked feature in regard to confirmation dialogs is the option to define completely custom buttons, which you can do by setting the `buttons` configuration of the dialog to an object structured like either one of the following samples:

Table 6.2 Configuration options related to buttons, handlers, and scoping of functions

Option	Description
`Buttons`	This option defines which button(s) to show in the message box. Out of the box four single button(s) are defined: ▪ `Ext.MessageBox.OK`: Shows a simple OK button ▪ `Ext.MessageBox.CANCEL`: Shows a simple Cancel button ▪ `Ext.MessageBox.YES`: Shows a simple Yes button ▪ `Ext.MessageBox.NO`: Shows a simple No button In addition to these four buttons there are three button sets provided: ▪ `Ext.MessageBox.OKCANCEL`: Shows an OK/Cancel button combination ▪ `Ext.MessageBox.YESNO`: Shows a Yes/No button combination ▪ `Ext.MessageBox.YESNOCANCEL`: Shows a Yes/No/Cancel button combination
`fn`	This option allows you to define a callback function that's invoked whenever one of the buttons from the `button` configuration is clicked. By default buttons provide their own `scope`, which means the callback wouldn't necessarily have access to the rest of your application. To alleviate this you can use the `scope` parameter.
`scope`	This optional parameter allows you to change the scope of the callback function to provide easy access to whichever level of code you need within the callback.

```
Ext.Msg.show({
    title   : 'Are you sure?',
    msg     : 'This will shutoff the internet!',
    fn      : callbackFunction,
    buttons : {
       text  : 'go!',
       ui    : 'action'
    }
});
```

or for multiple buttons:

```
Ext.Msg.show({
    title   : 'Are you sure?',
    msg     : 'This will shutoff the internet!',
    fn      : callbackFunction,
    buttons : [
       {
           text   : 'go!',
           ui     : 'action'
       },
       {
           text   : 'umm, no'
       }
    ]
});
```

With alert and confirm out of the way the only mystery left is how to handle user input in a message box. To learn more you'll explore the prompt method in the next section.

6.3.3 Requesting input from users

Figure 6.12 Basic message box with input prompt

Prompts provide a way to ask the user for a single line or multiple lines of text. Anything more than that and you'll have to look into building a form, something covered in chapter 7. Prompts build on top of the alert and confirm functionality we covered in the previous section. The title is retained and the message body is appropriated as the label for the input field. Instead of using a Yes/No button like the confirm box, prompts use an OK/Cancel button combination by default (see figure 6.12).

The default prompt uses a single-line input field, with autocomplete, autocapitalize, and autocorrect turned off. Just like the other helper functions, prompt calls the show method of the message box, setting which buttons you want to use. Showing a prompt follows almost the same syntax as the alert and confirm dialog showcased earlier:

```
Ext.Msg.prompt('Sorry dude!', 'What is your name again?', callbackFunction);
```

This code represents the standard call for a prompt. Notice that we said that it *almost* follows the same syntax. The major difference is the addition of extra parameters to set whether the input field should be multiline, whether it should contain an initial value, as well as a configuration object to turn on or off things like autocomplete or set the maximum length of the field. Showing a multiline prompt looks like this:

```
Ext.Msg.prompt(
    'Sorry dude!',
    'What is your name again?',
    callbackFunction,
    this,
    true,
    'Anthony',
    {
        maxlength      : 180,
        autocapitalize : true
    }
);
```

The first two parameters are the title and message body; parameters 3 and 4 are the callback function and its scope. Parameter 5 sets the prompt to be multiline. Setting `this` to `true` means the input field will utilize the default height of 75 pixels. You could provide a number instead of `true`, and the supplied number becomes the height of the input field. The next parameter is the initial value you want the input field to have, whereas the last parameter represents the configuration for the input field. Fully rendered, the prompt looks like figure 6.13.

That's all there is to prompts. The main thing to keep in mind is that you're not bound to use the `prompt` helper function. You can always call the `Ext.Msg.show` method directly and pass a configuration object with your desired combination of options. To recap the options related to text input take a look at table 6.3.

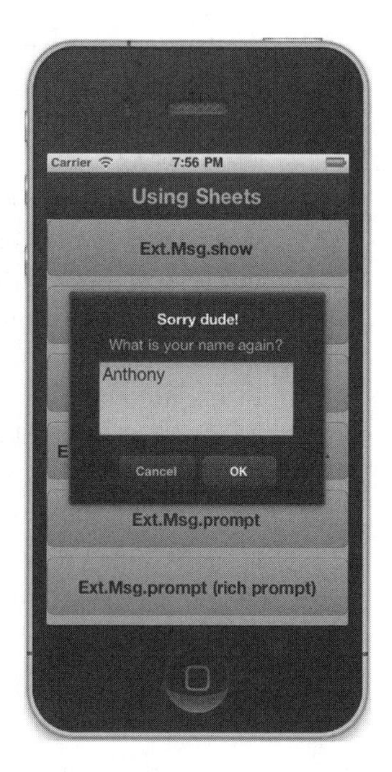

Figure 6.13 A multiline prompt

Table 6.3 Available configuration options for prompting users for input

Option	Description
`multiLine`	This option can either be a Boolean value or a number. In the event it's set to `true` the input box will be multiline and use the default 75-pixel height. In the event that you provide a number the number specified will be used as the height for the input box.
`prompt`	This option takes a configuration object that allows you to specify properties for the input field. The options include any of the available options for a `textfield` instance, which are `autocomplete`, `autocapitalize`, `autocorrect`, `maxlength`, `autofocus`, and `placeholder`. All of these have defaults that are automatically set, and you wouldn't need to bother with most of them.
`value`	Automatically populates the input box with the defined `value`. This option is only useful when you're prompting the user to provide input.

If you happen to look through the documentation, you might notice the `show` method specifies `cls`, `height`, `width`, and `defaultTextHeight` as valid options to be passed. The truth is that as of this writing these parameters don't accurately work; the `show` method ignores them. To use any of these options you'll have to instantiate your own `Ext.MessageBox` instance and pass this option into the config of the `MessageBox` (instead of the `show` method), and then call the `show` method of your custom `MessageBox` instance. Using the provided `Ext.Msg` singleton won't work with these options.

There you have it, ladies and gentlemen, `MessageBoxes` in a nutshell! A round of applause, please.

6.4 Summary

Your newfound knowledge of how to overlay information onto the current screen with sheets, how to notify and prompt the user for information via message boxes, and how to present users with an abundance of choices through use of pickers should help make your apps more intuitive and user-friendly.

You're starting to amass an impressive list of tools in your tool belt, ranging from managing components, to using panels and tabs, to docking items, to harassing the user with those pesky pop-up messages. Although this is an impressive list you might be thinking, "What we've learned so far is great, but I have all this data on a remote server and I don't know what to do with it!" You're absolutely correct. Without further ado, the next chapter will introduce you to the exciting world of data stores, lists, and the ever-so-powerful data view.

7 Data stores and views

This chapter covers

- How data stores work with models, proxies, and readers
- Using DataView
- Exploring lists
- Digging into nested lists

You've learned quite a lot so far about how to work with Panel and Sheet, and you can build a great-looking application. But the application you're able to build so far is static and may need to consume or save remote data. Say you're tasked with building a mobile viewer so that anyone in your client's company can search a global address book and find contact information while they're traveling. You'll likely want some sort of list that will load the contact list from a remote server. Building such a list might sound complicated, and it is, but Sencha Touch makes it easy to achieve.

In this chapter you'll explore Ext.data.Store and the classes it uses to send a request to a server and read the response. We'll also look at displaying this data in Ext.DataView.DataView and Ext.DataView.List, and you'll learn when to use them.

Before we jump into using the widgets you need to learn about data stores. A list can't display anything without a data store, so let's begin with an overview of what a data store is and how it works.

7.1 Examining data stores

`Ext.data.Store` is one of the most commonly used classes in Sencha Touch. From the perspective of someone just learning Sencha Touch it can be quite a daunting task to understand how the store works and consumes remote data. You're going to take it one step at a time. First, you'll build your knowledge of how a store sends a request to a server and decodes data so that the store can properly consume the data.

7.1.1 The anatomy of data stores

If you're familiar with databases, then an easy way to think of a store is like a table in a database; it has fields and it has rows of data. That's the simplified version, but the `Store` class has many different supporting classes that it uses to load remote data and read the data and consume it. To get data from a server-side source you use a proxy to send your request to the server. The server will then respond with a format such as JSON or XML and you'll use a reader to decode the data, allowing the store to read the response. Looking at figure 7.1, you can visualize how the process happens: the store only talks to the proxy, and the proxy does the coordinating. There are some other classes in the mix, but generally speaking this is a great overview of the workflow the store uses to retrieve remote data and work with it. To add some flexibility you can use a `Model` class that can hold the fields, use a proxy and reader, build associations with other `Models`, and even validate data. If you change a `Model` instance, or record, that's within a store and you want to send that change to the server, the store will use a writer to get the needed data and encode it for transfer by the proxy to the server. Figure 7.1 shows the main classes the store uses.

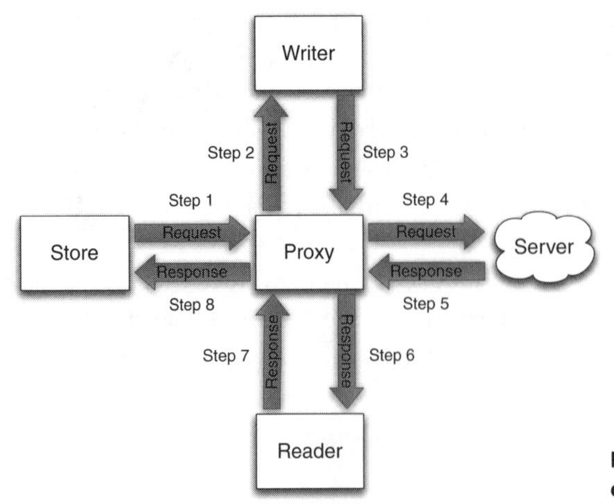

Figure 7.1 Overview of how the store can load remote data from the server

Figure 7.1 shows you an overview of the workflow for the store to load remote data in eight steps:

1. The store tells the proxy that it wants to make a request.
2. The proxy asks the writer if it has anything it can add to the request that the proxy will make. This is where changes you've made to the store, such as creating, editing, or removing a record from the store, will synchronize with the server.
3. If the writer has anything to add it'll add data to the request. The data can be encoded using JSON or XML and other options we'll discuss in a little bit.
4. The proxy now has everything it needs to make the request, so it sends the request to the server. The type of request depends on what proxy you're using. For instance, an Ajax proxy sends an Ajax request, but the `LocalStorage` proxy uses HTML5's `localStorage` property.
5. The server does whatever it needs, such as gathering data or handling the new, edited, or removed records, and responds. The response will come back in either JSON or XML, depending on what you chose as your response.
6. The proxy gets the response from the server but it's basically just a string, either JSON or XML, with JSON being recommended. As is, the store won't be able to handle this raw data, so the proxy sends the raw data to the reader for decoding.
7. The reader decodes the raw data, taking the JSON or XML and turning it into JavaScript.
8. The round-trip for the request is now complete. The proxy will create records from the JavaScript and return the records to the store.

Now the different widgets are able to display the data. It's good to know about `Ext.data.Operation`. This class does much of the heavy lifting behind the scenes and works in parallel with the proxy. This class is more for advanced users; you're not going to deal with `Operation` directly. Let's slow down and review the various classes that you just saw in more detail.

7.1.2 Using proxies to load data

As we mentioned, you use a proxy in a store to load data, but what does that mean? The most popular proxy class is the `Ext.data.proxy.Ajax` proxy, which uses an `Ext.Ajax` request to get data from a remote data source.

There are other proxies that you can use depending on your server requirements, but they all work in the same way. You have Ajax, Direct, JsonP, LocalStorage, Memory, Rest, and SessionStorage proxies, and they all work the same. They're just different ways of communicating with a data source. Table 7.1 shows some of the commonly used configurations of the proxies.

NOTE All proxies, except the default Memory proxy, load data asynchronously, so you need to use a callback or an event to take action when a store loads.

Table 7.1 The common configurations of a proxy

Config	Description	Defaults
`url`	The URL of the remote data source.	Empty string, `""`
`api`	An object of URLs for each CRUD action. You should only use this over the `url` config if you need to have different URLs for each action.	`{ create : undefined, read : undefined, update : undefined, destroy : undefined }`
`actionMethods`	An object of methods to use for each CRUD action.	`{ create : 'POST', read : 'GET', update : 'POST', destroy : 'POST' }`
`extraParams`	An object of parameters to include in the request. These will always be sent but can be overridden in the `params` config of the `load` method.	Empty object, `{}`
`reader`	A configuration object to configure the reader.	`{ type : 'json' }`
`writer`	A configuration object to configure the writer.	`{ type : 'json' }`

There are a couple of websites that can help you debug responses if you're returning JSON data. If you're using the Ajax proxy with the JSON reader, www.jsonlint.com is a great place to test the response coming from the server so that it's returning what you're expecting, and it'll also format the JSON so you can better read the response. If you're using the JsonP proxy, www.jsonplint.com is the site to do your server testing and response visualization. Both sites can handle remote requests, or you can paste the response to lint the data and its format.

Cross-domain loading

JSONP is a means of loading cross-origin content without CORS (cross-origin resource sharing) to get around browser security. CORS allows you to make Ajax calls cross-origin, but it requires server setup. You'll find a quick reference to setting up CORS at www.enable-cors.com/. The server has to respond with a valid JSONP response. A valid response is JSON surrounded by a callback function whose name is sent in the request. For example, if in your request you have a callback parameter with a value of `'Ext.data.JsonP.callback1'`, then the response should look like this: `Ext.data.JsonP.callback1({"foo": "bar"});`. Learn more at www.jsonplint.com.

7.1.3 *Using readers to digest data*

Like the different proxies, there are three readers you can use: array, JSON, and XML. It's recommended that your server return JSON; therefore, you should use

`Ext.data.reader.Json` to decode that JSON. If your server returns XML you should use `Ext.data.reader.Xml`. Table 7.2 shows some important configurations of a reader.

Table 7.2 Common configurations of a reader

Config	Description	Type	Defaults
`idProperty`	The field name in the response to use as the ID of each record.	String	`'id'`
`messageProperty`	The property in the top level of the response to use as the message of the response. This is useful if there was an error and the server needs to reply with the error message.	String	null
`rootProperty`	The property in the top level of the response to use as the root of the data.	String	Empty string, `''`
`successProperty`	The property in the top level of the response to use as the success signal.	String	`'success'`
`totalProperty`	The property in the top level of the response to use as the number of total results. This number is used for paging and tells the store the total number of results, not how many are in the returned page.	String	`'total'`

7.1.4 *Understanding models*

Models are the rows of data within the store and must have the fields defined to represent the value in each model. In a real-world application you should have your code organized so that you always have a model defined for a store to use; you shouldn't define fields on the store.

Models are records

In the Sencha Touch community we call records *instances* of models. We'll be using these two terms moving forward.

You know a store can have a proxy defined, but there's some flexibility when using models. You may need to load a single row of data, and using a store is a bit of overkill because it handles an array of data. To handle this use case the model can have a proxy defined so it can load that single row of data.

If you have a store that's using a model that has a proxy defined the store will then use the proxy defined on the model. If you have a proxy defined on both and you try to load the store the proxy defined on the store will then be used.

7.1.5 *Writer to sync*

The last bit we'll talk about to complete the store ecosystem is the writer. A writer's job is to include new, edited, or destroyed records in the sync request from the store. You execute a sync request by executing the `sync` method on the store, and if there are any dirty records they'll be included in the request. A "dirty record" is a record that's new, edited, or destroyed. You'll also see the term "phantom record," which is a new record on the client side that hasn't yet been synced with the server side.

As with the proxies and readers various writers are available depending on your situation. Table 7.3 shows the common configurations of the writers, but each writer has its own specialized config for its request format, so there are some others.

Table 7.3 Common configurations of a writer

Config	Description	Type	Defaults
`encode`	Only for JSON writer. Set to `true` to encode the request and be placed in the request body, or if `rootProperty` is set as the value of that property.	`Boolean`	`false`
`writeAllFields`	For both JSON and XML writers. Set to `true` to send all field values, else send only the ones that were changed.	`Boolean`	`true`
`rootProperty`	Only for JSON writer. The property in the top level of the request to send the encoded data.	`String`	`'records'`
`documentRoot`	Only for XML writer. The property to be used as the root of the XML document.	`String`	`'xmlData'`

7.1.6 *Simple store example*

The following listing shows a simple example of how to set up a local store. It's called "local" because the data is specified locally (in this case, inline), whereas a remote store would load remote data via a proxy.

Listing 7.1 Basic local store example

```
Ext.define('MyModel', {                          ← 1 Defines Model
    extend : 'Ext.data.Model',

    config : {
        fields : [                               ← 2 Specifies fields
            'name',
            { name : 'twitter', type : 'string' }
        ]
    }
});                                              3 Instantiates store
var store = Ext.create('Ext.data.Store', {       ←
    model : 'MyModel',
    data  : [                                    ← 4 Includes data
        { name : 'Mitchell Simoens', twitter : 'SenchaMitch' },
```

```
        { name : 'Jay Garcia',       twitter : 'ModusJeasus' },
        { name : 'Anthony De Moss',  twitter : 'ademoss1'    }
    ]
});
```

As you can see in listing 7.1, implementing a local store with inline data is simple. You're using something new, `Ext.define` ❶, which we'll cover more in chapter 10. `Ext.define` creates a new class definition, and in this case, a `Model` class to be used for the `Model` instances. A model has to have fields, so you use the `fields` configuration to specify an array of fields. You can use a string or a configuration object to specify your fields. If you specify a string as you did with `name` it'll be used as the name config and no conversion will happen. But if you specify a configuration object ❷ it'll be used to create the `Ext.data.Field` instance. If you need to do value conversion you can specify the type configuration; refer to `Ext.data.Field` for more configurations you may need.

Now you get into creating the store ❸. You use the model configuration to tell the store which `Model` class definition to use. To finish off a simple local store you give it inline data ❹, which is an array of objects. Notice the properties `name` and `twitter` in the data objects, which will map to the fields in the `MyModel` `Model` class definition. With this code your store now has three records (`Model` instances) and is ready for a view to display this data.

Now let's look at implementing a proxy and reader to use with a remote store. Configuration is pretty straightforward, but we'll highlight some of the main areas in the next listing. You'll reuse the `MyModel` model you created in listing 7.1.

Listing 7.2 Basic remote store example

```
var store = Ext.create('Ext.data.Store', {
    model    : 'MyModel',
    autoLoad : true,
    proxy    : {
        type   : 'ajax',
        url    : 'authors.php',
        reader : {
            type          : 'json',
            rootProperty  : 'authors',
            totalProperty : 'totalNumber'
        }
    }
});
```

If you remember that you're reusing the `MyModel` class from listing 7.1 and you picture the anatomy of the store from figure 7.1 it's easy to understand how this remote store works. By default, the store won't load automatically, so you can set the `autoLoad` configuration to `true` to make the store load right away. You could also execute the `load` method after the store was created to have control of when the store loads. To load remote data you specify a proxy to be used. You'll use Ajax to load remote data, so specify `type` and `url` so you can use the Ajax proxy. This may throw you, but looking back at figure 7.1 you can see the proxy talks to the reader, not the store, so you put the reader configuration in the proxy config object, not the store. To configure the reader tell it to use the JSON reader with the type configuration and also use the rootProperty and totalProperty to properly map where the data is and the total number of records in the database.

This Store configuration would expect the following response from authors.php:

```
{
    "success": true,
    "totalNumber": 3,
    "authors": [
        {
            "name": "Mitchell Simoens",
            "twitter": "msims84"
        },
        {
            "name": "Jay Garcia",
            "twitter": "_jdg"
        },
        {
            "name": "Anthony De Moss",
            "twitter": "ademoss1"
        }
    ]
}
```

You now know how to configure both local and remote stores and what a sample response would look like for the remote store. In chapter 8 we'll walk you through saving data back to the server. Now it's time for the fun part: learning about and creating a few different views to use the store to display data. First, let's look at the DataView.

7.2 Implementing DataView

You know how to get data from a data source, but now you need to learn how to display this data. For that you can use the Ext.DataView.DataView widget that takes the records from a store and displays them with an XTemplate. You can get as advanced as you like or keep it simple.

7.2.1 How DataViews work

Just as when we walked through the various classes for the store, we'll show you what classes DataView uses. You learned in chapter 3 that there's a Component model; one Component extends another Component in order to give stability and enforce standard behavior. DataView extends Container and each record in the store gets rendered as an item to DataView, so now you have items within the parent container (DataView). Figure 7.2 illustrates how DataView extends Container and gets the records from the

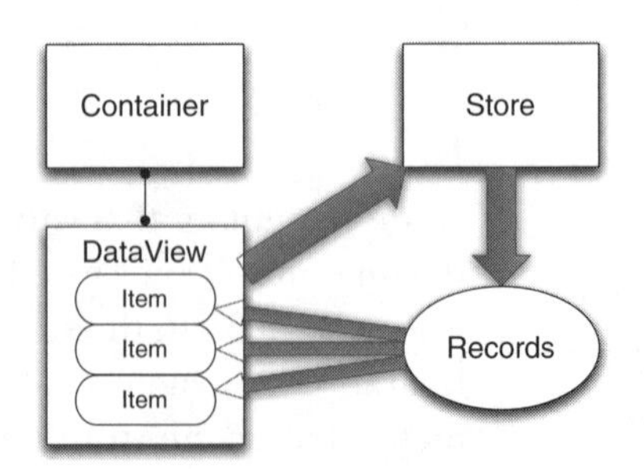

Figure 7.2 Showing the workflow of how DataView gets the records from the store to render its items

store to render its items. What does that mean? Well, DataView gets all the records within the store and loops through the array to generate HTML for each row using the record data and the template you defined in the itemTpl configuration. What's that itemTpl configuration? It's an Ext.XTemplate configuration, usually a string that generates HTML when you pass it data.

7.2.2 Walking through XTemplate

XTemplate is a subclass of Template. We like to think of an XTemplate as a template on steroids. It can map data in a basic form or it can get more advanced, where you can loop through an array in your data, use template methods to establish some logic to change or format what's to be displayed, and even execute JavaScript. The XTemplate is intimidating, we'll give you that. But by going over it piece by piece, we'll show you how to take advantage of this class easily.

In the next listing we'll look at the XTemplate's most basic form, where you can map data to display data-driven HTML.

Listing 7.3 XTemplate example

```
var template = Ext.create('Ext.XTemplate',
      'Book: {data}'
);

var html = template.apply({
     data : 'Sencha Touch in Action'
});

Ext.getBody().setHtml(html);
```

That was easy! First you created an XTemplate and passed in a couple parameters. In this case you passed in two strings, but you can pass in a couple of different things. You can pass in one string or a number of strings that'll be combined into one. You can also pass in an array that'll be joined into a single string. You then used the apply method on the XTemplate to apply the object of properties and values that'll get mapped. This object has a data property that'll get mapped to the {data} string within the string. Finally, you want to display the HTML that was just created, so you use the setHtml method on Ext.getBody(), which is just the document.body. The results of listing 7.3 can be seen in figure 7.3.

Figure 7.3 Results of the basic XTemplate

Our examples are becoming increasingly advanced, and you can see how it's not that difficult to take advantage of the power behind the XTemplate. Next we'll tackle looping through an array so you can display data that may be in array form. For this you use a `for` loop as we would in JavaScript, but you don't have to worry about setting an iteration variable or specifying a length so the `for` loop can know how to loop through an array; the XTemplate will take care of this for you, as shown in the following listing.

Listing 7.4 XTemplate array looping

```
var template = Ext.create('Ext.XTemplate',
    'Book: ',
    '{data}',
    '<tpl for="authors">',                                  // 1 Uses <tpl> tag to loop
        '<p style="padding-left: 2em;">{name}</p>',         // 2 Maps to name property
    '</tpl>'
);

var html = template.apply({
    data    : 'Sencha Touch in Action',
    authors : [                                             // 3 Specifies authors array
        {
            name : 'Mitchell Simoens'
        },
        {
            name : 'Jay Garcia'
        },
        {
            name : 'Anthony De Moss'
        }
    ]
});

Ext.getBody().setHtml(html);
```

The example in listing 7.4 builds on listing 7.3 to easily show how to loop through data. When the XTemplate parses the string you pass it, if it sees the `<tpl>` tag ❶ it knows that it has something to do. In this case you use the `for` attribute, which tells it that it needs to loop through the `authors` array in the data passed in. At each iteration through the `authors` array that object will be applied to the string within the `<tpl>` and `</tpl>` tags, so you can use `{name}` ❷ to map to the `name` property of the objects within the `authors` array ❸. It can sometimes be hard to visualize what this will produce; figure 7.4 shows the output. Notice the indented lines, which show what's in the array that you looped through.

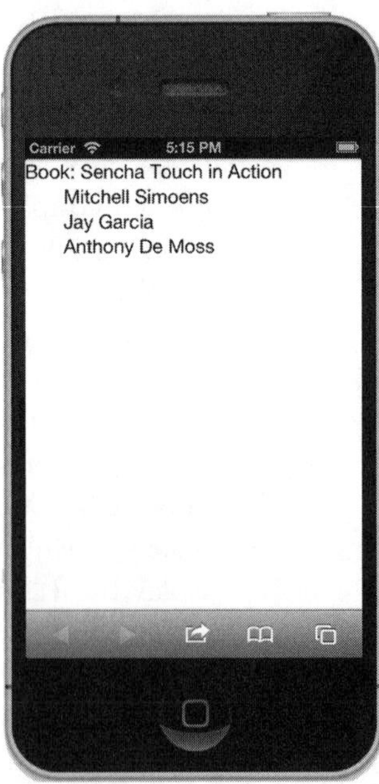

Figure 7.4 Results of looping through an array within an `XTemplate`

The next example shows two ways you can format the data that gets mapped. Say you're expecting a `Date` object and you want to format it. The next listing shows how you can use `Ext.util.Format` and also how you can use member functions to return a formatted date string.

Listing 7.5 Using methods within an XTemplate

```
var template = Ext.create('Ext.XTemplate',
    '<p>Ext.util.Format: ',
    '{dateVal:date("Y-m-d")}</p>',                       ❶ Passes in format string
    '<p>Member Function: ',
    '{[this.formatDate(values.dateVal)]}</p>',           ❷ Executes member function
    {
        formatDate : function(date) {                     ❸ Specifies member function
            return Ext.util.Format.date(date, "Y-m-d");
        }
    }
);

var html = template.apply({
    dateVal : new Date()
});

Ext.getBody().setHtml(html);
```

In this listing you see an example of how to define and use member functions for the XTemplate. When you see `{dateVal:date("Y-m-d")}` it means that you want to execute `Ext.util.Format.date`, passing in the value of the mapped data. In this case it's the `dateVal` property, and you also pass in the format string ❶. This is the same as doing it in the member function, which you tell it to execute via the square brackets ❷. The scope is that of the XTemplate, and the member functions specified in the object (in the same place you specify the strings) are placed on the XTemplate. You have to get the data point from the `values` object, which the `formatDate` member function ❸ will execute when you apply data to the template. Note that you should always return something in the member function. Figure 7.5 shows the output of listing 7.5 using the shortcut method ❷ and the member function ❸.

Figure 7.5 The output of two ways to execute functions within the XTemplate

The XTemplate isn't meant to replace JavaScript; it just makes it easier to have control over what's to be displayed by allowing JavaScript to be executed from within the XTemplate. The following listing shows how to execute JavaScript from within an XTemplate.

Listing 7.6 XTemplate array looping

```
window.STIA = {};                                          ❶ Creates STIA namespace

STIA.formatDate = function(date, format) {
    return Ext.util.Format.date(date, format);
};                                                         ❷ Creates function to execute

var template = Ext.create('Ext.XTemplate',
    '<p>Javascript execution: ',
    '{[STIA.formatDate(values.dateVal, "Y-m-d")]}</p>'     ❸ Executes JavaScript function
);

var html = template.apply({
    dateVal : new Date()
});

Ext.getBody().setHtml(html);
```

As listing 7.6 shows, executing JavaScript from within an XTemplate is similar to executing member functions. First you specify the function that you want to execute; in this case, it's a `formatDate` method on the `STIA` namespace ❷ that you created ❶. Then just as you did with the member function, you execute the method within the square brackets ❸. The results of listing 7.6 are shown in figure 7.6.

We've walked through some useful features of XTemplates to display data-driven HTML. You may be thinking the XTemplate is great, and it is, but it has a downside. With everything you do in a mobile app, you have to think about performance. If you overuse XTemplates rendering DataViews may be very slow because all the code within the XTemplates will be executed for each record in the store. So if you have 50 records in the store your code will execute 50 times. You need to ensure your XTemplate is as small as you can make it while still accomplishing what you need to.

Figure 7.6 The output when executing your own JavaScript within an XTemplate

7.2.3 *Implementing your first DataView*

You've learned about the store and its proxy, reader, and model; you've also learned how the `DataView` class works and the functionality of the XTemplate. How do you put all that together to make a fully working DataView? That's what we're going to look at next, and it's easier than you may think.

You'll use the store from listing 7.1 to have the data present for your DataView to display, and you'll use the remote store in listing 7.2 when you try out the list. First let's look at the next listing to see how you can implement your first DataView.

Listing 7.7 Basic DataView

```
var dataview = Ext.create('Ext.dataview.DataView', {
    fullscreen : true,
    store      : store,
    itemTpl    : [
        '{name} ',
        '<a href="http://www.twitter.com/{twitter}" target="_blank">',
            '@{twitter}',
        '</a>'
    ]
});
```

The first step is to instantiate the DataView, in this case via direct instantiation. You use the `fullscreen` configuration to add the DataView to `Ext.Viewport` in order to render the DataView to the DOM so you can see the data. You use the store from listing 7.1 with inline data by using the `store` configuration. Lastly, you use the XTemplate to get the data from the store to display using the `itemTpl`. Because the store has data in it DataView will take that data and apply it to the XTemplate you specified in the `itemTpl` configuration, which will result in figure 7.7. You use `{name}` and `{twitter}` to tell the XTemplate where to display the data from the store.

One thing to know about `itemTpl` is that it can accept a few different types of values. In listing 7.7 you specified an array of strings, but you could've specified a string or an actual `XTemplate` instance. It's common to just specify a string or an array of strings; we used an array only to fit it on the page of this book.

You just learned about DataViews, which can display data from a store. Now let's talk about a subclass of `DataView: List`.

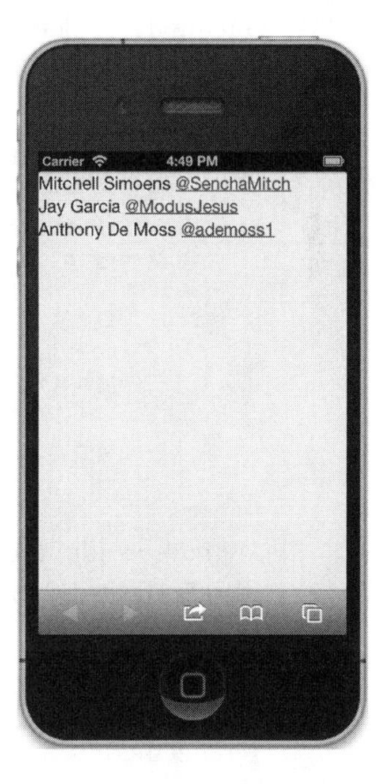

Figure 7.7 Results from listing 7.7 showing inline data from the store using the DataView

7.3 Advanced features with List

So far in this chapter you've learned how to get data with the store and display the data in the DataView using the XTemplate. In this section you'll look at the `Ext.dataview.List` component and learn to use its extra functionality.

7.3.1 How List differs from DataView

The `List` class extends `DataView`, so you get all the functionality that you've learned about so far in this chapter, such as getting data from a store and displaying it with an XTemplate. There are several things that `List` can do that `DataView` can't:

- *CSS*—`List` has some CSS that `DataView` doesn't have.
- *Infinite*—`List` can handle an infinite amount of data in the store. This is sometimes known as an *infinite scrolling List* and is only available in Sencha Touch 2.1 and later.
- `IndexBar`—This is a vertically stacked alphabet that when tapped on will jump to a group of records. You should only use this when your list is grouped.
- *Group data*—Each of your records can be grouped and have a header to specify the group.
- *Disclosure icon*—An icon typically appears to the right of each row that signifies that an action can take place when you tap on that icon.

Taking it slow, let's see how to create a basic list. In the next listing you'll use the same store and `itemTpl` you did in listing 7.7 so that you can compare the results of the basic list and the basic DataView.

Listing 7.8 Basic list

```
var list = Ext.create('Ext.dataview.List', {          <- 1 Creates list
    fullscreen : true,
    store      : store,
    itemTpl    : [
        '{name} ',
        '<a href="http://www.twitter.com/{twitter}" target="_blank">',
            '{twitter}',
        '</a>'
    ]
});
```

We have to admit something: we used a common copy/paste shortcut that you should avoid. All we did in listing 7.8 was change the first line ❶. For this basic list the configurations are exactly the same as the basic DataView in listing 7.7. Let's look at figure 7.8 to see the differences.

On the left we have the DataView and on the right we have the list. The main difference here is visual, because the list has some CSS to differentiate the rows. Another difference is if you tap on a row in each the list displays the row with a blue background and white text to signal it has been selected, whereas the DataView doesn't. You'd need to handle that in your CSS.

7.3.2 CSS differences between List and DataView

As you can see in figure 7.8, the list has a different look than the DataView. The structure of the DOM (Document Object Model) or the elements that you see is much the same, so why do they look different? It's because the CSS that's bundled with Sencha Touch has styling for `List`, but `DataView` is meant to allow developers to style it how they need without overriding styles. Another CSS difference is visible when a row is selected. Both `DataView` and `List` can allow each row to be selected, and if you look at the CSS classes on the row element you'll notice it gets a CSS class name added, usually `x-item-selected`. When you select a row in the DataView visually nothing will happen, but if you select a row in the list you'll see that row will get a blue gradient

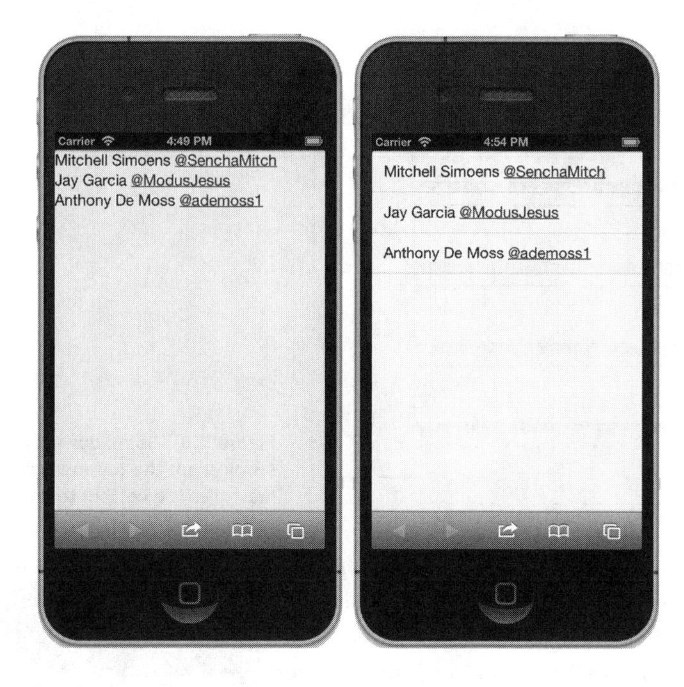

Figure 7.8 Differences between a basic DataView (left) and basic list (right)

background and the text will be white to show the selection. You could specify your own CSS to handle selection, but with `List` you'll have to keep in mind that there's already CSS styling that you may have to override; with `DataView` there's none.

CSS styling is the first of the differences between `List` and `DataView`. The other three are just more advanced features, so they're going to be grouped into one section and example. It'll be exciting to see more visual changes happen, but before you can play you must first do your homework.

7.3.3 *Using infinite data with List*

New in Sencha Touch 2.1, `Ext.dataview.List` now supports infinite data. Now you can have 100 `Models` in your store and performance will still be high just as if you had only 10. In fact, you can have thousands in your store and it'll perform very high. This is a major update in Sencha Touch 2.1. Let's see how it works and whether there are any negative side effects.

Unlike `DataView`, `List` doesn't have a 1-to-1 relationship between a `Model` instance and a DOM node. Instead, `List` figures out how many rows can fit in the screen, adds some for a buffer, and reuses them on the fly as you're scrolling. So say your screen can fit 10 rows on screen; `List` will then use 15 rows so there's a little buffer as you're scrolling, because it takes time to change the data in a row. So you have a total of 15 rows rendered and you scroll down; the first couple of rows are transformed to the bottom using CSS and data from the next `Model` instances is applied to the rows. As you scroll up or down, data is constantly being moved and written to these rows. This is how the list can be infinite: it's reusing 15 rows for possibly hundreds or thousands of rows. Figure 7.9 illustrates this row movement.

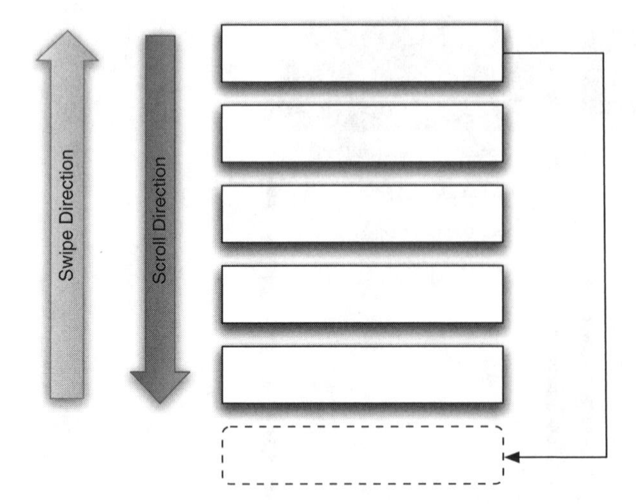

Figure 7.9 Scroll down by swiping up; the top row will be moved to the bottom to be reused.

This sounds great, but all of this moving and writing can come at a cost. In iOS we've found the scrolling to be smooth. Android has always been plagued with subpar CSS transform performance except in the Android 4.1 and 4.2 versions and especially when using Chrome for Android. In older or underpowered devices the scrolling may be a little twitchy. To combat this, the only thing you can do is keep the `itemTpl` template as minimal as possible. That way, there aren't many DOM elements being created, removed, and transformed.

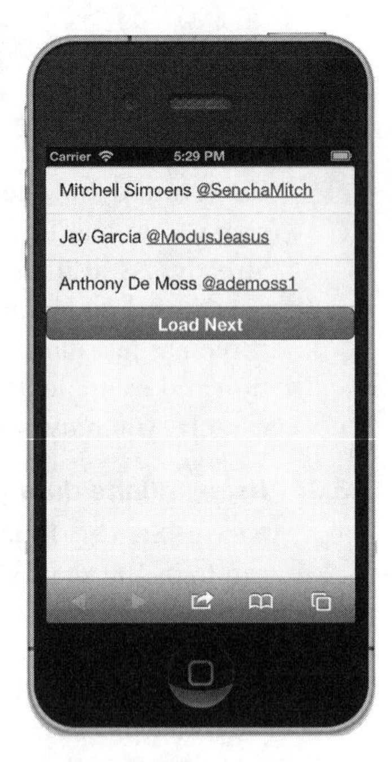

Figure 7.10 Show a button under a list

Another functionality that's lost is making a list unscrollable. Because the list doesn't have a 1-to-1 relationship between a `Model` instance and a DOM node, not all models in a store are going to be rendered. So it's not feasible to have an unscrollable list or else you won't be able to see the models that aren't rendered; it depends on the scrolling. You may be asking yourself why this matters; some use cases require a button or some other component at the bottom of the list so you can take some action when you scroll to the bottom, as shown in figure 7.10. To achieve that in Sencha Touch 2.0 nest a list within a container and make the container, not the list, scrollable. That way, you can have the list and button as child items of the container, and because the container is the component that's scrolling you'll see the button at the bottom of the container. You can see something like that in the following listing.

Listing 7.9 Button under a list

```
Ext.create('Ext.Container', {                          <-- 1 Creates wrapping container
  fullscreen : true,
  scrollable : 'vertical',                             <-- 2 Sets container to scroll vertically
  items      : [
    {
      xtype      : 'list',
      scrollable : false,                              <-- 3 Disables scroll on list
      itemTpl    : ''.concat(
        '{name} ',
        '<a href="http://www.twitter.com/{twitter}" target="_blank">',
          '@{twitter}',
        '</a>'
      ),
      store      : {
        fields  : ['name', 'twitter'],
        data    : [
          { name : 'Mitchell Simoens', twitter : 'SenchaMitch' },
          { name : 'Jay Garcia',       twitter : 'ModusJeasus' },
          { name : 'Anthony De Moss',  twitter : 'ademoss1'    }
        ]
      }
    },
    {
      xtype : 'button',                                <-- 4 Shows button
      text  : 'Load Next',
      ui    : 'confirm'
    }
  ]
});
```

In listing 7.9 you create a container ❶ and set `scrollable` to `'vertical'` ❷. For the child list you disable the scrolling by setting `scrollable` to `false` ❸. The container has a second child, a button ❹ that'll show under the list. The list will be the height that the rows take up and the button will be right under the last row in the list. This means that if you had a long list the container would control the scrolling so that once you got to the end of the list you'd see the button. Remember, this code will work with Sencha Touch 2.0 but not 2.1 and newer.

In Sencha Touch 2.1 and later the list has to be scrollable, so this technique won't work. To support this functionality you can have components docked to the list, but it's not as easy as setting the `docked` config on the component. If you set the `docked` config the component will be docked and won't participate in the scroll, which isn't exactly what you want to happen. A new config has been created called `scrollDock` that tells the infinite scrolling list that this item is to be docked to the top or bottom when the scroll has gone to the top or bottom. Along with the `docked` config you can now accomplish what you want, which is to see the component at the bottom. The following listing shows the `docked` and `scrollDock` configs in use.

Listing 7.10 New button under list

```
Ext.create('Ext.dataview.List', {                          <-- 1 Creates list
  fullscreen : true,
  itemTpl    : ''.concat(
    '{name} ',
    '<a href="http://www.twitter.com/{twitter}" target="_blank">',
      '@{twitter}',
    '</a>'
  ),
  store      : {
    fields : ['name', 'twitter'],
    data   : [
      { name : 'Mitchell Simoens', twitter : 'SenchaMitch' },
      { name : 'Jay Garcia',       twitter : 'ModusJeasus' },
      { name : 'Anthony De Moss',  twitter : 'ademoss1'    }
    ]
  },
  items      : [                                           2 Creates button as item of list
    {
      xtype      : 'button',                               <--
      docked     : 'bottom',                               <-- 3 Docks button under last list item
      scrollDock : 'bottom',                <-- 4 Docks to bottom
      text       : 'Load Next',
      ui         : 'confirm'
    }
  ]
});
```

Although a few differences exist between listing 7.9 and 7.10 both show the same thing, as seen in figure 7.10. Because you don't need to wrap the list in a container you create a `List` instance ❶. You create a button as an item of the list ❷ and give it the `docked` config ❸ so it will display after the last list item. If you didn't use this config the button would display above the last list item. You also set the `scrollDock` config ❹ to the same value as `docked` so that the docking of the button will happen when the scrolling of the list is at the bottom.

7.3.4 Advanced features for List

There are three more differences between `List` and its superclass `DataView`. Because they're features we can save time by discussing each and then seeing a single example. Don't worry; we'll walk through the code so you can see which lines are doing what to accomplish what you see in figure 7.11.

INDEXBAR

Many native apps that list data have what's called an IndexBar. In figure 7.11 the IndexBar is on the right side of the list. It's a vertical stack of the alphabet docked to the right side of the list, and when tapped it allows your list to jump down to the group that matches the letter tapped. You'll usually only see the IndexBar when a list is grouped or sorted. Also note that it'll show all letters, even if, as in figure 7.11, there are only three groups.

GROUPED DATA

Say you have a list of people who have both a first name and a last name, as people usually do. You could present the application user with a long list of people, and maybe even sort this list by last name. You can imagine what this list would look like. Our users may get lost scrolling through a long list searching for a particular person. A better solution is to provide your users with a header over a group of records. For instance, all last names that begin with A will be under the A header and last names that begin with X will be under the X header. Looking at figure 7.11 you can see a short list grouped with the blue headers over each group. If you were to scroll and have more rows in the list the header would be docked to the top until the next header reached the top, where that header would then be docked to the top. The store will need to be configured to be able to group (which will also sort based on the grouping) so that the data being provided to the list is in order, which makes it easy for the list to display the data. We'll look at this store modification shortly.

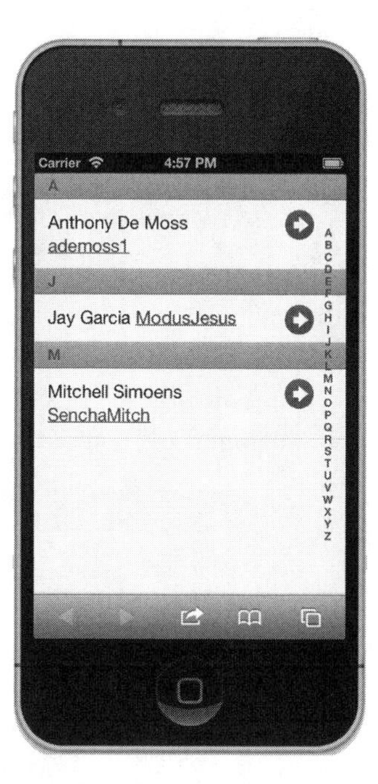

Figure 7.11 The list showing the IndexBar, groups, and disclosure icons

DISCLOSURE ICON

Still looking at figure 7.11, notice the icons on each row floating to the right. These icons, called disclosure icons, aren't just cosmetic but will change and add behavior when tapped. Normally when you tap on a row it'll be selected, but if you tap on the disclosure icon it won't select the row but will instead fire the `disclose` event on the list. The `onItemDisclosure` config on the list can be set to a Boolean value or a function. If `onItemDisclosure` is specified as a function, when you tap on the disclosure icon it'll fire the `disclose` event and will also execute the function, giving you two ways to handle the disclosure icon tapping. In your application we suggest sticking with the `disclose` event, but in some cases it may be easier just to handle the tap in the `onItemDisclosure` config function. In our upcoming example you'll use the `disclose` event and specify the `onItemDisclosure` config as the function, but know that you could just specify the `onItemDisclosure` configuration as `true` and use the `disclose` event.

We just went over three new features of the list that the DataView doesn't have. You learned that the list has some CSS styling that the DataView doesn't have. Can you imagine what all these features look like? No? Well, let's build an example and see what the list will look like with all these features working together.

7.3.5 Example of IndexBar, grouping, and disclosures

In listing 7.11 you'll enable the IndexBar, grouping, and disclosures to show how configurable the list is and how easy these features are to include. You'll also use the remote store from listing 7.2 with a little modification for grouping so that you can see what happens when the list has to wait for data to load so it can display. You can add the following grouper configuration to the store to enable grouping:

```
grouper  : {
    groupFn : function(record) {
        return record.get('name')[0];
    }
}
```

For each record this gets the value from the `name` property and returns the first letter. This now groups all records on the first letter of the `name` property. Now let's put it all together in an advanced list in the next listing.

Listing 7.11 Advanced list

```
var list = Ext.create('Ext.dataview.List', {
    fullscreen         : true,
    store              : store,
    indexBar           : true,                          ❶ Uses IndexBar
    grouped            : true,                          ❷ Enables list grouping
    onItemDisclosure   : true,                          ❸ Shows disclosure icons
    itemTpl            : [
        '{name} ',
        '<a href="http://www.twitter.com/{twitter}" target="_blank">',
            '{twitter}',
        '</a>'
    ],
    listeners          : {
        disclose : function(list, record) {             ❹ Listens to disclose event
            Ext.Msg.alert(
                'Author Tap',
                'You tapped on ' + record.get('name')
            );
        }
    }
});
```

First, you set the `indexBar` config ❶ to `true`. You enable grouping using the `grouped` config ❷, so now you have the headers to distinguish the groups. To display the disclosure icons you set the `onItemDisclosure` config ❸ to `true` and listen for the `disclose` event ❹ to show an `Ext.Msg.alert` when you tap on one of the disclosure icons.

While the list and DataView are loading the list will by default show an instance of `Ext.LoadMask` (figure 7.12, on the left). You can configure this by using the `loadingText` config of `List` or `DataView`. The middle image in figure 7.12 is what the list looks like with the IndexBar, grouping, and disclosure icons all enabled on the list. The IndexBar is the vertical alphabet on the right. If you had a long list and tapped on a letter the list would scroll down to the beginning of that group. If a group doesn't exist for that letter the list will scroll to the letter before it. To show that the list is grouped it displays the headers for each group with the `groupFn` value in the bright blue header. Do you see the blue circle with a white arrow in it? That's the disclosure

Figure 7.12 The list loading (left), the three features (middle), and handling the `disclose` event (right)

icon, and if you tap it listing 7.11 will display an `Ext.Msg.alert` when the `disclose` event fires (figure 7.12, on the right).

You can use each of these three features of `List` independently of one another, so you can have just disclosure icons or grouped data with or without the IndexBar. This example shows an advanced example of what `List` is capable of out of the box.

Now you've seen the features of the list, and things look really good! You can build some fantastic, visually rich lists for your users. Next we're going to talk about `NestedList`, which is an extension of `Container` that uses child lists to display hierarchical data in a card layout.

7.4 Displaying hierarchical data with NestedList

If you have flat data you should use `DataView` or `List`. But if you have hierarchical data then neither is equipped to handle that data. They'll show the first level, also called the root level, of data, but not any of the child data levels. Not to leave you out in the cold, Sencha Touch has a component equipped to handle hierarchical data called the nested list. As the name suggests, the nested list component will show a list for each nested level of data. The `NestedList` class extends `Container` and uses the card layout. Each item is a list and the number of items depends on the number of levels in your hierarchical data. The nested list also has a top-docked title bar to show a back button when not on the root level, allowing you to traverse backward. This title bar will also show the title text, usually the same text that's shown on the row you tapped on.

To get the data you don't just use a regular store like you used for `DataView` and `List`; `Store` also doesn't support hierarchical data. For this you have to use `TreeStore`.

7.4.1 *Understanding the hierarchical data*

Before we dive into `NestedList` and `TreeStore` let's see how the hierarchical data should look. Hierarchical data is nested data where each record is known as a node. The data becomes hierarchical when a node contains child nodes, and those can contain child nodes, and so on. A node that doesn't have any children is known as a leaf. You're likely familiar with the file system on a computer where folders can have folders and files. A node that has children is like a folder and a leaf is like a file.

Here's what a basic JSON sample of hierarchical data looks like:

```
{
    "children" : [
        {
            "text"     : "Mitchell Simoens",
            "children" : [
                {
                    "text" : "@msims84",
                    "leaf" : true
                },
                {
                    "text" : "http://www.linkedin.com/in/mitchellsimoens",
                    "leaf" : true
                }
            ]
        },
        {
            "text"     : "Jay Garcia",
            "children" : [
                {
                    "text" : "@ModusJesus",
                    "leaf" : true
                },
                {
                    "text" : "http://www.linkedin.com/in/tdginnovations",
                    "leaf" : true
                }
            ]
        },
        {
            "text"     : "Anthony De Moss",
            "children" : [
                {
                    "text" : "@ademoss1",
                    "leaf" : true
                },
                {
                    "text" : "http://www.linkedin.com/in/ademoss",
                    "leaf" : true
                }
```

```
                ]
            }
        ]
}
```

Looking at this JSON you can see the hierarchy. The top level is the root of the data, which has a single property, named `children`. Each level will have a `children` property to specify what children that item has, unless that item has the `leaf` property. It's assumed each item has children unless the `leaf` property is set to `true`, meaning that it won't have children; it's the end of the road. Each item, leaf or not, will have a `text` property that'll be displayed.

7.4.2 Using TreeStore

As we mentioned at the beginning of section 7.4, you need to use a TreeStore, not a regular store, to consume hierarchical data. There isn't a difference when creating a TreeStore, but the TreeStore is able to handle hierarchical data whereas a store can't.

Using the same JSON you saw in section 7.4.1, the following listing is a sample TreeStore with the associated `Model`.

Listing 7.12 Sample TreeStore

```
Ext.define('Author', {                                    <-- ❶ Creates Author Model
    extend : 'Ext.data.Model',

    config : {
        fields : [
            'text',
            'link'                                        <-- ❷ Specifies Model fields
        ]
    }
});

var store = Ext.create('Ext.data.TreeStore', {            <-- ❸ Creates TreeStore
    model    : 'Author',
    autoLoad : true,
    proxy    : {
        type  : 'ajax',
        url   : 'authors-tree.json'
    }
});
```

Creating a TreeStore is easy. As with a regular store, you create a `Model` class definition; in this case you're naming it `Author` ❶. The only thing required is that you specify which field you intend to display, which in this case is text. You also specify another field that you'll use later, named link ❷. You then create the TreeStore ❸ and give it the `model`, `autoLoad`, and `proxy` configs, and within the `proxy` config you're going to use the Ajax proxy with `url` set to the authors-tree.json file that contains the JSON from section 7.4.1.

Run the code from listing 7.12 and the TreeStore will load the JSON from the `authors-`tree.json file and parse the hierarchical data to be used in a nested list.

7.4.3 *Creating a basic nested list*

So you can load data, but that's no fun without being able to display the data. As we discussed earlier, you can't use a DataView or a list to display hierarchical data. For this task Sencha Touch has the nested list component. As mentioned in section 7.4, the nested list is just a container that uses the card layout to display child lists. The following listing is a basic example of how to display the JSON from section 7.4.1 using the TreeStore from listing 7.6.

Listing 7.13 Basic NestedList

```
var nestedlist = Ext.create('Ext.dataview.NestedList', {
    fullscreen : true,
    store      : store,
    title      : 'Authors'
});
```

There's not much going on in this simple nested list. You create the nested list and give it the `fullscreen` configuration to add it to `Ext.Viewport` so it can be displayed. You also specify the store for the nested list to use. You give it the `title` config, which is a title for the root level. The `title` config is optional, but without it the nested list will look a little funny, because the top-docked title bar will be blank. With `title`, when you are on the root level the title bar will display this value so your users can tell what they're looking at. Check out figure 7.13 to see what you get when you run listing 7.13.

As you can see in figure 7.13, the image on the left is the root level. You can see the title that you specified in listing 7.13 at the top. Without it you'd just get the blue

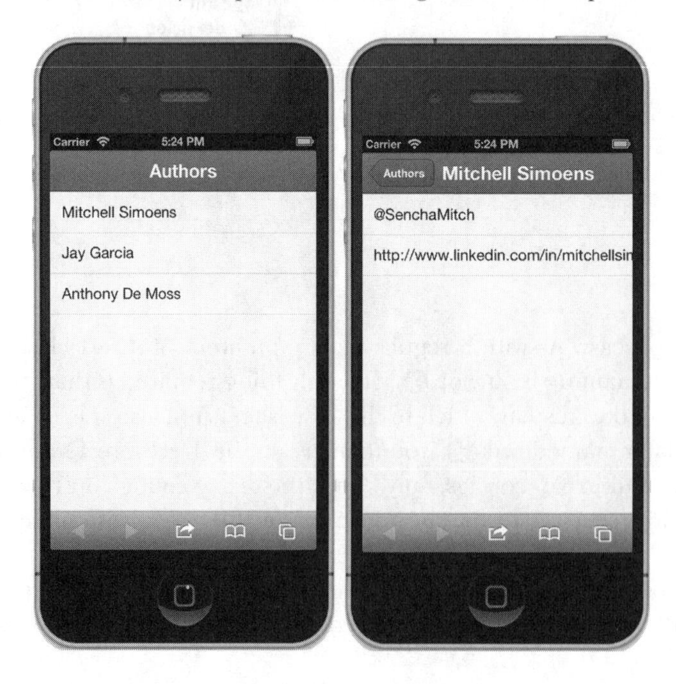

Figure 7.13 Results of listing 7.13 showing a simple NestedList

toolbar docked there, which doesn't look so great. Plus, with `title` set, you can now see that this list is named Authors. If you tap on the first list item, Mitchell Simoens, the image on the right would animate to show the children of the Mitchell Simoens list item. Note that the top title bar now displays the same text for the title as the list item you tapped on. If you tap on Anthony De Moss, the title bar would say Anthony De Moss instead of Mitchell Simoens. If the text were longer than could fit in the title bar the text would be cut off and an ellipsis used to give a better presentation rather than just letting the text run off the screen. Also note in the title bar that you now have a button that says you can go back to the Authors list. This button uses the back `ui` config and uses the text from what was in the title bar as the title. In this case the last title was Authors. If you didn't specify the `title` config on the nested list the button would display the text Back, so the back button should never be blank.

7.4.4 Showing details

So what happens when you get to the level that's a leaf? You can tap on it and it'll be selected, but nothing happens. Because it's a leaf there are no more children, but usually you'll want to show details about that leaf. A configuration of the nested list that you haven't covered so far is the `detailCard` config. Tapping on a leaf will create and add a detail card to the nested list and animate to this component. The detail card can be another list, a container, or a form—just about anything you need it to be.

Let's take a look at how to work with `detailCard` in the next listing.

Listing 7.14 Nested list details

```
var nestedlist = Ext.create('Ext.dataview.NestedList', {
    fullscreen : true,
    store      : store,
    title      : 'Authors',
    detailCard : {                                          // ❶ Specifies config detailCard
        xtype : 'container',                                // ❷ Uses Xtype container
        tpl   : [                                           // ❸ Includes template to display data
            'Titter Page: ',
            '<a href="{link}" target="_blank">',
                '{text}',
            '</a>',
            '<br /><br />',
            'Tapping on this link will take you outside of this app'
        ]
    },
    listeners  : {
        leafitemtap : function(nestedlist, list, index, t, record) {   // ❹ Adds listener to leafitemtap
            var detailCard = nestedlist.getDetailCard();    // ❺ Gets detailCard instance

            detailCard.setData(record.getData());           // ❻ Applies data to detailCard
        }
    }
});
```

You took the nested list from listing 7.14 and added `detailCard` ❶. This is the configuration object that'll be used to create the component that will display details about the leaf that was tapped. This detail card will use the XType `'container'` ❷ and a template using the `tpl` configuration ❸ to accept data and display it. If you left it here you'd get a blank screen, so you need to add a `leafitemtap` event listener ❹ that'll only be fired when you tap on a list item that's a leaf. Within this `leafitemtap` event listener you need to get the reference to the detail card using the `getDetailCard` method ❺ on the nested list, which is passed as an argument on the function. Finally, to show data you execute the `setData` method on the `detailCard` instance, passing in the data from the record that was tapped on ❻. Now you should see what's shown in figure 7.14.

Figure 7.14 View of the detail card in the nested list

Instead of another list you'll see the component you specified in the `detailCard` config. With the logic in the `leafitemtap` event listener you applied the data from the record you tapped on to the detail card. Now you have the ability to create a view that can handle hierarchical data and even show a detail view on it. Can you imagine what you can use `NestedList` for? It's powerful but simple to configure.

7.5 Summary

In this chapter, we covered how to use data and what views to use to build more dynamic applications. You first learned how to use the store and its proxy, reader, writer, and model classes. You also saw how to use these classes to build powerful, data-driven applications.

You then learned how a DataView works and what an XTemplate is. You implemented some common configurations in order to display data from a store. This was your first step into a data-driven view that you'll more than likely need to use in your applications.

Next you built on the DataView with the subclass `List`. You saw the differences between `DataView` and `List` and then implemented an IndexBar, grouped your list, and learned what a disclosure is and how to take action on a user tapping on the disclosure icon. You learned that a list can visually and functionally be an important component to have in your arsenal.

Lastly, you learned what hierarchical data looks like and how to consume it with the specialized TreeStore. You used the nested list component to display this data one level at a time and used the `detailCard` config in `NestedList` to display details about a leaf item.

In the next chapter we'll enter the world of forms, with all of its various fields, and you'll learn how to submit data from the form to a remote server.

8 Working with forms

This chapter covers

- Building simple and complex forms
- Loading and saving forms
- Binding a form to a list

So far in this book we've covered many ways to manage data and display it onscreen using Sencha Touch's various UI widgets, ranging from tabs, to generic components, to XTemplates. Though this information is important for most applications, we're sure your users are just as demanding as ours and want to interact with the data—and maybe even save it!

Interacting with the data is done using Form panels and form fields. In this chapter you'll examine the various form elements, their uses, and how to put them together in a form that can be loaded and saved. After that you'll explore ways to use forms in conjunction with other components, like the list sample from chapter 5. So what are you waiting for? Let's get started.

8.1 What makes Form panels so special anyway?

It should come as no surprise that developing and designing forms represents one of the common tasks in every app developer's life, and as such, we developers are a

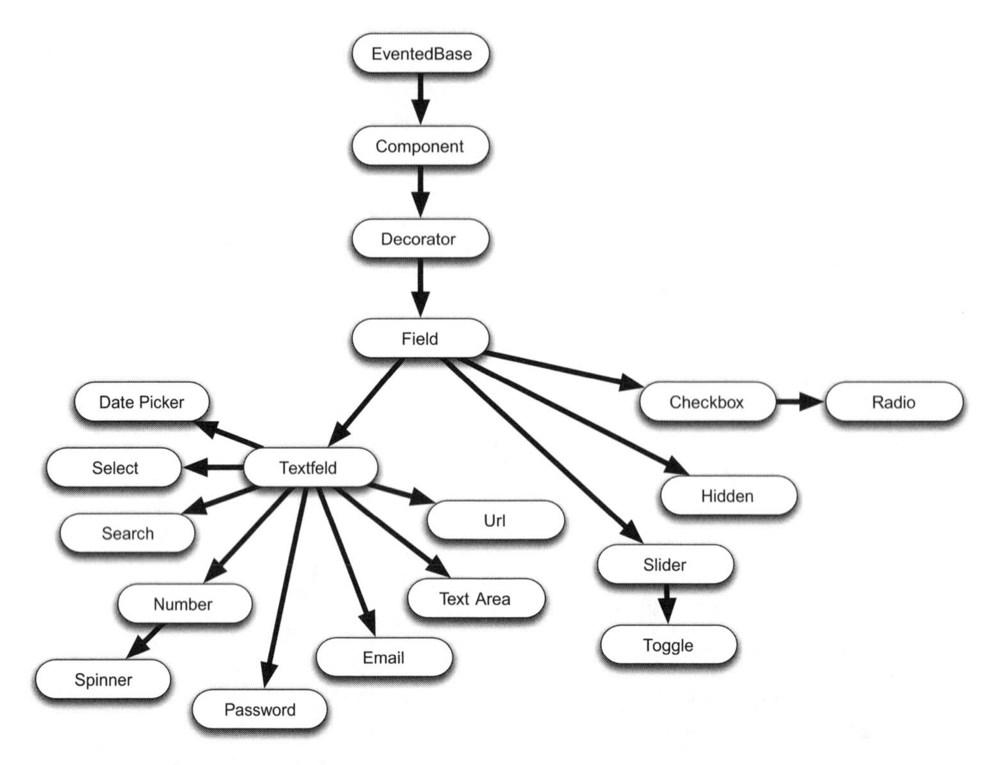

Figure 8.1 Forms-related widgets and their inheritance

demanding bunch. A way to load the form? You bet. A way to save it? Of course. Validation? An absolute must!

With the bar set high it's only natural for Sencha to tap into the extensive set of widgets developed for Ext JS, Sencha Touch's bigger cousin, and bring some of those features over to provide you with an impressive list of different Form elements. Figure 8.1 gives an overview of all the forms-related UI widgets and their inheritance chain.

Forms in Sencha Touch are rendered using the `Form Panel` class, which gets filled with one of the various UI widgets that require input. These range from generic text fields, to date pickers, to sliders, all of which extend the `Field` class, as shown in figure 8.1.

`Form Panel` follows our familiar Component model pattern and extends the base `Panel` class, as shown in figure 8.2. As such, it has all the standard capabilities of a panel, like layouts, ways to manage children, and a few additional methods to handle submitting and loading of form data.

The `Field` class in turn extends the `Decorator` class, which provides extra visual decoration around the base `Component` class and provides functionality to render labels next to the form fields, get and set the value of a form field, or even mask the field during load operations.

Back in chapter 3 we talked about the Component model and the stability and predictability it introduces to the framework. The relationship between the `Form Panel` and `Field` classes is a prime example to highlight this paradigm. The `Field` class exposes several functions to allow getting, setting, or resetting the value of the field. In addition, it tracks whether the field is "dirty," which refers to whether the field value has changed since the initial load. The `Form Panel` understands these things about every field intrinsically and uses these methods to automatically populate any of its child fields whenever the form is loaded or submitted. This is possible because each field instance exposes these same functions, creating the aforementioned predictability.

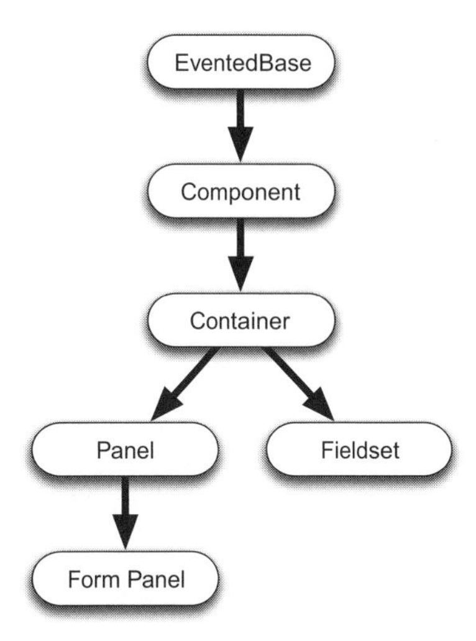

Figure 8.2 Inheritance chain for `Form Panel` and `Fieldset`

> **"Sencha Touch" !== "Ext JS"**
> If you're coming over from the Ex JS side of things you'll find that forms in Sencha Touch deviate slightly from their Ext JS variant. These deviations range from a drastically different, dare we say prettier, visual design, to a more simplistic layout that better accommodates touch input. Under the hood are various changes as well. As of this writing there's no built-in validation for forms, and the load method that was your bread and butter for populating forms with data has been replaced with a methodology that forces you to use models instead.

Now that you have a basic overview of forms and their relationship to fields it's time to put this knowledge into practice and build a form.

8.2 Building a basic form

Like all the other `Container` subclasses the `Form Panel` class can use any layout that's available in the framework to create exquisitely laid-out forms. Before delving into multicolumn or multitabbed forms, something that requires you to go through quite a few design and usability considerations, let's review the basics and use the default `VBox` layout to stack form elements one on top of another.

In the next listing you'll put together a simple form that can be used to edit data.

Listing 8.1 A basic form

```
var myForm = Ext.create('Ext.form.FormPanel', {        // 1 Instantiates Form Panel
    fullscreen : true,
    items      : [
        {   xtype        : 'textfield',                // 2 Instantiates text field
```

```
            label       : 'First',
            name        : 'firstName',
            placeHolder : 'Enter First Name Here'          <-- 3 Assigns placeholder text
        },
        {
            xtype       : 'textfield',
            label       : 'Last',
            name        : 'lastName',
            placeHolder : 'Enter Last Name Here'
        },
        {
            xtype       : 'selectfield',
            label       : 'Status',
            name        : 'inviteStatus',
            placeholder : 'nothing',
            options     : [                                <-- 4 Defines options for select field
                {
                    text  : 'Undecided',
                    value : 'undecided'
                },
                {
                    text  : 'Accepted',
                    value : 'accept'
                },
                {
                    text  : 'Declined',
                    value : 'decline'
                }
            ]
        }
    ]
});
```

The code should be pretty straightforward by now. You start by creating a `Form Panel` instance ❶ and giving it the `fullscreen` option. Doing so works because `Form Panel` is an extension of the `Panel` class, and as such the `fullscreen` option will make it occupy the entire screen. Note that without this option the form wouldn't show on the screen, because Sencha Touch always requires one container to take up the entire screen by using the `fullscreen` option. For the child elements of the form you define two text fields ❷, one for first name and one for last name. You give each text field placeholder text ❸ that'll automatically be shown whenever the text field is empty. In general it's best to try to make the placeholder text as useful and instructive as possible to give users an idea of what they're supposed to do on your form. Lastly, you define a select field and give it three options ❹ that, depending on the platform on which this example is running, will show up either as an ActionSheet or a drop-down when the field is activated. Once you execute the code from listing 8.1 you should get a result that looks like figure 8.3.

In case you're wondering about the one property we didn't talk about yet, the `name` property, it serves one purpose: to provide an identifier for each field, which the `Form Panel` uses whenever you load data into or retrieve data from the form.

Figure 8.3 Basic form sample on the left and the same form with text fields filled in and a select field activated on the right

During load operations a JSON object containing name/value pairs is used. The name of each property in the JSON object is mapped to a field that shares the same name within the `Form Panel`. The form automatically handles assigning the values from the JSON object to each respective field. A sample JSON object you could use to load data into the form from listing 8.1 looks like this:

```
{
    firstName    : 'Anthony',
    lastName     : 'De Moss',
    inviteStatus : 'accept'
}
```

On the flip side, retrieving values works almost exactly in reverse. The form exposes a convenient `getValues` method that returns a JavaScript object containing name/value pairs for each field in the form. Running the following line in the Safari debug console would return a JavaScript object that looks exactly like the one you used for loading:

```
myForm.getValues();
```

You're probably wondering just exactly how loading and submitting of form data works. Don't fret; we'll cover this topic in greater detail in later parts of this chapter, after you've gotten a better understanding of all the various form elements and the caveats to keep in mind about each.

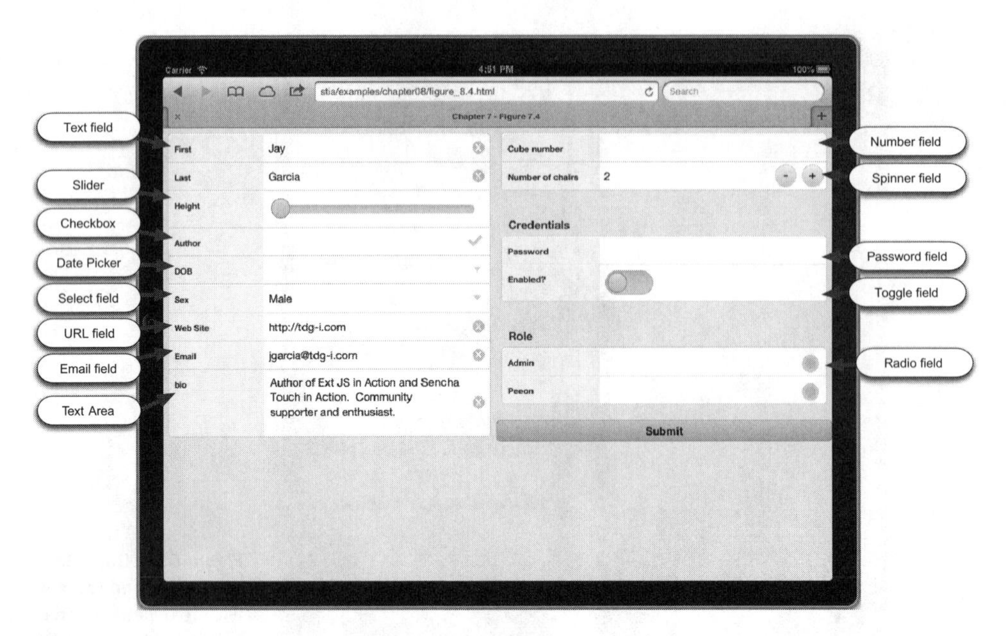

Figure 8.4 Overview of various form widgets

8.3 An overview of the different Form widgets

Sencha Touch offers quite a few wrapped native HTML5 input fields, as well as a few custom widgets. The fields text field, check box, URL field, email field, text area, number field, password field, and radio all implement native HTML5 input elements with additional styling to make everything look the "Sencha way." Each one of these widgets, with the exception of the radio and check box fields, will force the native slide-in keyboard to appear when focused, thus allowing users to enter data into the field. Figure 8.4 illustrates what most of these input fields look like when rendered on the screen.

Now that you're equipped with some visuals let's explore each one of these form elements in more detail.

8.3.1 Text field

Text fields provide the base for quite a few of the other fields on this list. A text field wraps the standard HTML5 input field and exposes a set of convenient config options to customize the behavior to some degree or another. Instantiation is straightforward and looks like this:

```
{
    xtype          : 'textfield',
    label          : 'Text Field',
    placeHolder    : 'Something goes here',
    value          : 'This is a text field',
    required       : true,
    autoCorrect    : true,
    autoCapitalize: true,
    autoComplete   : true,
```

```
    maxLength     : 5
}
```

The label provides a convenient way to mark what your field is all about, and it's automatically attached to the left side of your field. Each and every form element can have a label associated with it. Keep in mind that the device you're using limits the size of a label. On a phone, for example, your labels are significantly smaller than on a tablet device, causing long text in the label to automatically wrap, which can look quite ugly.

All the text-based fields (URL field, email field, text area, and select field) provide for a placeholder text that's automatically shown whenever the field is empty. The placeholder can be used to augment the label and provide additional instructions to the user. Generally it isn't advisable to forgo the label in lieu of a placeholder because the placeholder disappears once a field is filled out, creating a scenario where the user has to guess the purpose of the field.

The `required` parameter may be confusing, especially for those developers coming from Ext JS. As of this writing Sencha Touch doesn't provide any built-in form validation, so this parameter simply marks a field as required by adding an asterisk to the label. There's no additional visual feedback if a field is left blank or to prevent a form from being submitted.

The `maxLength` parameter allows you to limit the number of characters that can be entered into a text field. Once the limit has been reached any additional characters typed on the keyboard are discarded. If you manually set the value of a text field through the `value` config parameter or the `setValue` function the `maxLength` parameter is circumvented and it's possible to add an arbitrary number of characters to the field.

The last three config options deal with autocorrection, autocapitalization, and autocomplete. The device you're using determines the usefulness of these options. On an iOS device (iPhone/iPad) autocapitalization will automatically enable the Shift key for the first character, and autocorrection will trigger the annoying correction bubble that shows up to suggest a "better" spelling for your word.

8.3.2 URL field

`Urlfield` extends the basic `textfield` and brings with it all the same configuration options. Under the hood the HTML5 input field is set to type `url`. The only real purpose this serves is that, depending on your device, the keyboard will automatically be put into "URL" mode, showing keys that are more useful for entering a URL when editing the field. From an implementation perspective, the code looks like the following:

```
{
    xtype : 'urlfield',
    label : 'Url',
    value : 'http://www.senchatouchinaction.com'
}
```

Because the URL field is based on the text field you could've used any of the other config options you used for the text field. Note that this field doesn't provide any validation to ensure that the URL is properly formatted.

8.3.3 Email field

The email field acts as a close cousin to the URL field and follows the same principle, except for email addresses. Whenever the field is edited the onscreen keyboard is switched to "email" mode to make it easier to enter an email address. As with the URL field you could use any other config option from the text field. Follow this code to instantiate an email field:

```
{
    xtype : 'emailfield',
    label : 'Email',
    value : 'youremail@yourcompany.com'
}
```

Just like the URL field, the email field doesn't provide any validation to ensure that the email address is valid or even that it's formatted correctly.

8.3.4 Number field

The number field is yet another field similar to the email and URL fields. It extends `textfield` and provides additional config options for minimum and maximum values, as well as a step value. At this point these options are only relevant if you plan on using your app in a nonmobile browser. Mobile browsers will render this field as a plain text field that brings up the numeric keyboard when the field is edited. Instantiation follows almost the same code as the text field before and looks like this:

```
{
    xtype     : 'numberfield',
    label     : 'Number',
    minValue  : 1,
    maxValue  : 10,
    stepValue : 1,
    value     : 7
}
```

Because HTML5 still isn't properly implemented in most browsers as of this writing, especially as far as field validations are concerned, it'd normally be possible to enter any arbitrary value into the field, instead of being restricted to only numbers. Fortunately for us, Sencha Touch comes to the rescue and does basic validation to ensure the entered value is indeed a number.

8.3.5 Password field

The password field is where you start seeing your first deviation from the text field. The field automatically obscures any user-entered text as asterisk characters. The same is true for manually assigned values, either through the `value` config option or the `setValue` function. Both of those will be obfuscated as well:

```
{
    xtype : 'passwordfield',
    label : 'Password',
    value : 'abc'
}
```

One of the most useful aspects of the password field is the fact that it allows for the same config options as the text field, including the `placeholder` option, which shows up as plain text instead of starred-out characters.

8.3.6 Text area

The text area is similar to the text field; the main difference is that the text area allows for multiline text, which is something the text field doesn't. Newline characters ("\n") are automatically interpreted by the field and will insert a new line into the text. Implementation follows the same principle as the text field and looks just as easy:

```
{
     xtype    : 'textareafield',
     label    : 'Text Area',
     maxRows : 5,
     value: "This is a larger text area.\n\nWe can even get multiple lines in
     here"
}
```

Notice the `maxRows` config option; it can be used to limit the number of rows you want the user to be able to enter.

8.3.7 Check box field

The check box field works similarly to its native-web counterpart, except it's stylized via Sencha Touch's own check icon to mimic native application behavior. Instantiation is easy and largely consists of only a label. It's possible to set the initial value of the check box via the `checked` option:

```
{
     xtype    : 'checkbox',
     label    : 'checkbox - ready',
     value    : 'yes',
    checked : true
}
```

The `value` option provides you with a way to set the value that'll be submitted if the check box is checked and a form submit occurs. In the previous example the value "yes" would be submitted, instead of the default "true" value.

8.3.8 Radio field

The radio field follows in the footsteps of the check box field, both visually and under the hood. It extends the check box and in fact looks the same once rendered. The only difference between it and the check box is the behavior when a radio field is activated: it switches between the different grouped fields, allowing only one of them to be checked at a time. To see an example of this take a look at the following code:

```
{
    xtype : 'radio',
    name  : 'myradio',
    label : 'radio - one'
},
```

```
{
    xtype : 'radio',
    name  : 'myradio',
    label : 'radio - two'
}
```

This code provides you with two radio fields that are automatically grouped together (not visually, but from a selection perspective). The grouping happens based on the name of the field. Any radio field that shares the same name with an already existing radio field is automatically added to the group. Whenever you click one of the two fields the other one becomes unchecked.

8.3.9 Date Picker field

The Date Picker field gives your users the ability to choose a date from a set by mimicking the native iOS Date Picker input widget. The Date Picker field implements a sheet, which is an overlay that slides in from the bottom, allowing the user to select values via vertical swipe or "flick" gestures. No matter what the device or its orientation the Date Picker field will always display a sheet, forcing selection though this modal overlay. The following code illustrates how to instantiate a Date Picker field:

```
{
    xtype : 'datepickerfield',
    label : 'Pick a Date',
    value : {
        year  : 2011,
        month : 2,
        day   : 23
    }

}
```

It's possible to set the initial value of the field via the `value` config option or the `setValue` method. Both ways accept either a JavaScript date object in the form of `new Date()` or a custom JavaScript object containing a year, month, and day property like the sample code earlier.

8.3.10 Spinner field

The spinner field is a custom-styled input field, allowing users to enter numeric values, much like the number field, with the addition of easy-to-use decrement and increment buttons on the side of the field. Notice that the increment and decrement buttons are larger than standard HTML spinner buttons, making them easier to use on a touch device. Implementation follows largely the same parameters as the number field:

```
{
    xtype      : 'spinnerfield',
    label      : 'Spinner',
    minValue   : 1,
```

```
    maxValue  : 10,
    increment : 2,
    cycle     : true,
    value     : 9
}
```

The main difference in config options compared to the number field is the addition of a `cycle` property. This causes the value to "wrap" around if `minValue` or `maxValue` is reached. For example, assume the current value is 9, the max is 10, and you have an increment of 2. The next click to increase the value would bring it over 10, thus causing it to wrap around to the bottom and start over at 1. Unlike the number field, the spinner field doesn't allow the user to enter values directly but instead forces a value change through the increment and decrement buttons.

8.3.11 Slider field

The slider field implements native Sencha Touch `Draggable` and `Droppable` classes, allowing users to input a numeric value via swipe and tap gestures. The slider can be restricted via the `minValue` and `maxValue` options, as well as the `increment` option. In code that looks like this:

```
{
    xtype     : 'sliderfield',
    label     : 'Slide me',
    minValue  : 20,
    maxValue  : 100,
    value     : 50,
    increment : 5
}
```

The `minValue` and `maxValue` don't change how far the slider can move but only change what the left edge and right edge represent respectively. The `increment` option changes where the slider snaps to when moved. This option takes into account where you drop the slider and automatically rounds to the nearest number that represents the next increment. One of the major drawbacks of the slider is the lack of visual feedback about its current value.

8.3.12 Toggle field

`togglefield` extends `sliderfield`, allowing users to toggle the field like an on/off switch via a swipe and tap gesture. Under the hood, the toggle field is just a slider with a `minValue` of 0 and a `maxValue` of 1:

```
{
     xtype : 'togglefield',
     label : 'Toggle Me',
     value : 1
}
```

You can set the initial state of the `togglefield` to on or off by setting the `value` field to `1` or `0` respectively.

8.3.13 Select field

The select field is the closest thing to a traditional combo box/drop-down you'll find in the Sencha Touch arsenal. It allows you to present the user with a list of choices to pick from, and whichever the user selects will be shown in the field. Depending on the device the select field displays different input widgets. On a phone it uses a sheet that slides in from the bottom, similarly to the Date Picker, and on a tablet it uses a small dialog-type control containing the options to select from. Here's an example implementation of a select field using hard-coded options:

```
{
    xtype   : 'selectfield',
    label   : 'Select',
    options : [
        {
            text  : 'Yes',
            value : 'yes'
        },
        {
            text  : 'No',
            value : 'no'
        }
    ]
}
```

In this case the defined options never change. Each option must have a text and a value field, where the `text` property represents the text that's shown in the list and the `value` property represents the value that'll be submitted when the form is submitted.

Quite frequently you'll find yourself in a situation where your data isn't formatted to contain a text or value field or where hard-coding the options isn't feasible. For those instances the select field provides additional options in the form of a `displayField` and a `valueField`. These are strings that determine which property from your data should be used instead of the `text` and `value` properties.

To fill your select field with dynamic data you can employ a data store, which triggers a remote load. In practice it looks like the code in the following listing.

Listing 8.2 Dynamically loaded select field

```
Ext.define('Person', {
    extend : 'Ext.data.Model',          <-- (1) Defines model with two fields
    config : {
        fields : [
            { name : 'name' },
            { name : 'pID' }
        ]
    }
});

var peopleStore = Ext.create('Ext.data.Store', {     (2) Uses model for store
    model    : 'Person',                              <--
    autoLoad : true,
```

```
    proxy     : {
        type    : 'ajax',
        url     : 'getPersons.json',
        reader  : {
            type           : 'json',
            rootProperty   : 'data'
        }
    }
});

var myPanel = Ext.create('Ext.form.Panel', {
    fullscreen : true,
    items      : [
        {
            xtype   : 'toolbar',
            title   : 'Tell us about yourself.',
            docked  : 'top'
        },
        {
            xtype          : 'selectfield',
            label          : 'Select Name',
            valueField     : 'pID',
            displayField   : 'name',
            store          : peopleStore
        }
    ]
});
```

❸ Sets URL of store

❹ Sets valueField and displayField

A lot happens in listing 8.2 just to load data into the select field from a remote source. You start by defining a model ❶ with two fields. In this case you're representing people, so you give everyone a `name` and a `pID` (short for personal ID) that'll serve as a unique identifier for each record. Of course, you need to put your model to good use, so you continue with a store that points to your model ❷. The store alone won't know how to get to the data. Because of that you need to define a proxy that tells the store from where and how the data is going to be loaded. The *where* is handled through the `url` property ❸, which points to a static JSON file. The *how* is handled through the reader, defined on the proxy, that uses a `type` of JSON, which just so happens to be the format of your data. Use the following dummy data and save it in a file named `getPersons.json`:

```
{
    "success" : true,
    "data"    : [
        {
            "name" : "Jay Garcia",
            "pID"  : 1
        },
        {
            "name" : "Anthony De Moss",
            "pID"  : 2
        },
        {
            "name" : "Mitchell Simoens",
```

```
                "pID"   : 3
            }
        ]
}
```

After you're done setting up the model and store you move on to create a Form panel with a select field and point it at the store you just created. You already know that your data won't have the standard `text` and `value` properties that the select field expects so you employ the `valueField` ❹ and the `displayField` properties and point them to `pID` and `name` respectively. This way, `name` is shown in the actual field and the `pID` would be used if you were to submit the form. That's all there is to loading a store dynamically.

By now you should have a good overview of the different form elements and how to go about creating a simple form as well as dynamically loading content into a select field. Obviously, nothing is ever simple in life, and the likelihood of being able to accomplish everything you need with a plain and basic form is virtually nonexistent. This is where some of the advanced techniques like multicolumn forms and fieldsets from the next section will come into play.

8.4 Building complex forms

Most of the forms you've built so far all have one thing in common: they follow the most basic design possible, stacking form elements in a single column, one on top of the other. This paradigm has certain advantages, especially when considering the fact that your form most likely has to work on multiple different devices, such as a variety of phones or tablets. For that reason a single column is the perfect fit because it works across the spectrum of devices. This doesn't mean that your form has to be one long and boring list of stacked elements, though. Sencha Touch provides for an easy way to split up your form visually by grouping elements into separate child containers with titles and instructions via the use of a fieldset.

8.4.1 More organized forms with fieldsets

The `Fieldset` class extends the `Container` class, and as such you can think of fieldsets simply as containers with a title. In practice the result of using a fieldset looks like the right side of figure 8.5, versus the standard stacked form on the left.

The way the form is broken up into smaller pieces should make the benefits of using a fieldset immediately apparent. Smaller pieces are more easily digestible by the end user and thus provide improved usability and flow. Place the following code within the `items` array of a Form panel to see it in action for yourself:

```
{
    xtype        : 'fieldset',
    title        : 'Personal Info',
    instructions : 'Enter your personal information',
    items        : [
        {
            xtype       : 'textfield',
```

```
            label       : 'First',
            name        : 'firstname',
            placeHolder : 'Enter your first name'
        },
        {
            xtype       : 'textfield',
            label       : 'Last',
            name        : 'lastname',
            placeHolder : 'Enter your last name'
        }
    ]
}
```

The code follows most of the samples we've already covered. You start by defining the fieldset using an `XType` and give it a title and instructions. Based on figure 8.5 you can see that the title shows up above the first field, and the instructions are shown below the last field. Although figure 8.5 doesn't illustrate this perfectly, instructions are centered and automatically wrapped to additional lines if necessary. Because the fieldset is itself a container it has its own `items` array where you place the form elements you wish to include. In this case you have two text fields with a label and placeholder text.

8.4.2 Multicolumn forms

While we've established that using fieldsets can dramatically improve the visuals of your forms, we still have one major problem: any tablet device wastes a sizable amount of screen real estate on wide fields when all the fields are organized in a single column. For such purposes, keep in mind that forms (and fieldsets) are simply containers that

Figure 8.5 A form without fieldsets on the left; the same form with fieldsets and instructions on the right

can contain their own layout. Knowing this you can create multicolumn forms that are specifically tailored to tablet devices with larger screens.

> **Mobile phones are cramped with tiny screens**
>
> It's important to remember that no matter how big the screen is on a mobile phone it's always going to be cramped and have less screen real estate than a tablet device. Consequently, multicolumn layouts are rarely suitable for phone devices, something that should be evident when taking a closer look at how cramped the labels are in figure 8.5.

The next sample ventures into the world of multicolumn forms. Take a peek at figure 8.6 to see what you're building. Note that this sample is made specifically for tablet devices and won't work properly on mobile phones due to space constraints.

Although all four of the fieldsets needed aren't terribly complex you'll start out by first defining each one separately before assembling it all together into a full Form panel at the end. The first two fieldsets in the next listing represent the left column, which consists of two text fields and a text area.

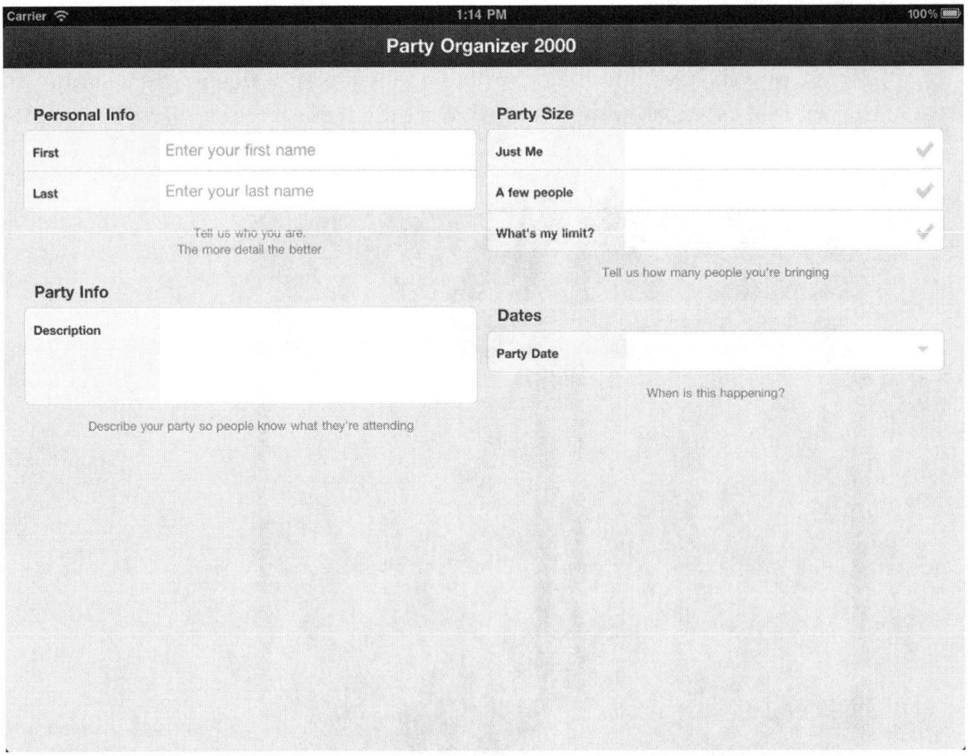

Figure 8.6 The multicolumn form with fieldsets you're going to build

Listing 8.3 Creating the fieldsets for the left column

```
var fieldset1 = {                                              ❶ Sets XType
    xtype        : 'fieldset',                                    to fieldset
    title        : 'Personal Info',
    instructions : 'Tell us who you are.'                      ❷ Provides title,
                        + '<br />The more detail the better',     instructions
    items        : [
        {
            xtype       : 'textfield',
            label       : 'First',
            name        : 'firstname',
            placeHolder : 'Enter your first name'              ❸ Defines
        },                                                        text fields
        {
            xtype       : 'textfield',
            label       : 'Last',
            name        : 'lastname',
            placeHolder : 'Enter your last name'
        }
    ]
};                                                             ❹ Defines second
var fieldset2 = {                                                 fieldset
    xtype        : 'fieldset',
    title        : 'Party Info',
    instructions : 'Describe your party so people know what'
                    + ' they are attending',
    items        : {
        xtype : 'textareafield',
        label : 'Description'
    }
};
```

For the most part you keep it short and simple by defining both fieldsets via `XType` configuration ❶ and providing a title and instructions ❷ for each fieldset. Notice that the instructions on the first fieldset contain an HTML line break to force the instructions into two lines. You could easily use any other HTML markup here to mark text in bold, italics, or underlined, or pepper it with CSS classes. You then proceed to populate the first fieldset with two text fields ❸, each of which receives a label, as well as placeholder text to provide even more ways to tell the user what to do. The second fieldset ❹ follows the same basic steps and simply contains a barebones text area.

From here you'll continue into both fieldsets in the right column, which you can see in the following listing.

Listing 8.4 Creating the second column of fieldsets

```
var fieldset3 = {                                              ❶ Adds three
    xtype        : 'fieldset',                                    radio fields
    title        : 'Party Size',
    instructions : 'Tell us how many people you\'re bringing',
    items        : [
        {
```

```
                xtype : 'radiofield',
                name  : 'size',
                label : 'Just Me',
                value : 'small'
            },
            {
                xtype : 'radiofield',
                name  : 'size',
                label : 'A few people',
                value : 'medium'
            },
            {
                xtype : 'radiofield',
                name  : 'size',
                label : 'What\'s my limit?',
                value : 'large'
            }
        ]
};

var fieldset4 = {
    xtype        : 'fieldset',
    title        : 'Dates',
    instructions : 'When is this happening?',
    items        : {
        xtype : 'datepickerfield',
        label : 'Party Date',
        name  : 'partydate'
    }
};
```

❷ Shares same name

❸ Configures Date Picker

As before, listing 8.4 starts out with the initial fieldset ❶, which you populate with three radio fields. Notice that each one of the radio fields has the same name ❷ in order to mark them as a set. Last but not least, you define the final fieldset ❸ and give it a Date Picker field that starts out blank.

This marks the last step in setting up all the components required to generate both columns. You're now ready to put it all together in the next listing and construct the Form panel that'll house everything.

Listing 8.5 Putting it all together

```
var myPanel = Ext.create('Ext.form.Panel', {
    fullscreen : true,
    scrollable : 'vertical',
    layout     : {
        type  : 'hbox',
        align : 'stretch'
    },
    defaults   : {
        flex  : 1,
        style : 'padding: 5px;'
    },
    items      : [
        {
```

❶ Starts with Form panel

❷ Uses HBox layout for columns

❸ Sets default values for columns

```
            xtype : 'toolbar',
            title  : 'Party Organizer 2000',
            docked : 'top'
        },
        {                                              ❹ Defines container
            xtype : 'container',                          for each column
            items : [
                fieldset1,                             Places fieldsets
                fieldset2                            ❺ in container
            ]
        },
        {
            xtype : 'container',
            items : [
                fieldset3,
                fieldset4
            ]
        }
    ]
});
```

Here you're finally creating your Form panel ❶ and putting everything together. Because the Form panel is nothing more than a special container you start by deviating from the standard `VBox` layout and give the Form panel an `HBox` layout ❷ with an `align:stretch` option instead. If you think back to chapter 4, the `HBox` layout provides you with a way to have multiple columns in a container and the `align` option is used for the vertical alignment, which in the case of `stretch` overrides the height, forcing each column to occupy the full height of the Form panel. You uniformly size each column horizontally through the use of the `flex` property, which you stick into a `defaults` object ❸ so it gets automatically applied to each column. In addition, you don't want the two columns to touch each other, so you set an additional padding of 5 pixels via the `style` property.

Next you move on to defining the actual containers ❹, which are downright easy compared to everything else. An `XType` definition as well as an `items` array that contains the fieldsets ❺ you want in each column does the trick. Each container will automatically use the default `VBox` layout and stack the fieldsets one on top of the other. In the end you come out with a form that looks like figure 8.6.

8.4.3 Doing more with your multicolumn form

Before we move on to loading content and managing data within forms let's chat about some of the possibilities we didn't cover. The form you just built is multicolumn through the use of the `HBox` layout and containers. Instead of doing that you could've designed the form differently and used fieldsets that span the entire width of the form but contain two columns within each fieldset. Because a fieldset extends the `Container` class it can take any of the layouts we covered in chapter 4, and as such you could define an `HBox` layout within a fieldset and then populate it with two containers, each containing the form components you want in each column. The possible combinations you could achieve are virtually endless. The one caveat you always have to keep in mind is that you should never nest Form panels within each other. There's no benefit to it, and in fact it causes tremendous problems that are hard to debug. If you do require multiple Form panels on a screen make sure to keep them separate.

By now you've seen how combining multiple components, fieldsets, and layouts can result in something that's both usable and space-saving. With that out of the way it's time we actually inject some function into form and talk about ways to load and submit data using forms.

8.5 Managing data with models

Submitting and loading data is one of the most crucial parts of forms, and incidentally one of the areas where new developers commonly get tripped up. Sencha Touch form submission requires a bit of rethinking from the old school, page-refresh-type form submission a lot of developers are used to. For those developers coming from the Ext JS side form submission will be a walk in the park. On the flip side, loading a form has changed quite dramatically and will most likely require getting used to.

8.5.1 Submitting data

Submitting a form requires only a few steps you need to be aware of. The first is that you need to get a handle on your Form panel. Use one of the component query methods we discussed in chapter 5 or assign the form to a variable the way you did in listing 8.5. The second thing you need to understand is how the `submit` method works. For that, take a look at the following listing, which is based on the code from listing 8.5.

Listing 8.6 Submitting your form

```
var formSubmit = function () {
    omyForm.submit({
        url     : 'success.json',
        success : function (form, response) {
            Ext.Msg.alert('Success', response.msg);
        },
        failure : function (form, response) {
            Ext.Msg.alert('Failure', response.msg);
        }
    });
};
```

In listing 8.6 you create a `formSubmit` function that handles submission of your form and that could be called from anywhere else in the code. Your handler function uses the `form.submit` method to take care of sending the data. The format is pretty much the same as it is in Ext JS; it consists of a `url` that determines where the form should be submitted, as well as a `success` and a `failure` handler.

Although you specify the URL in the `submit` call be aware that you could've just as easily used the `url` config option on the Form panel, which is automatically used when the `submit` method is called. The only reason you didn't do that in this sample is to illustrate the fact that the URL can be changed at runtime.

The success and failure handlers are callback functions called if the form submission was successful or failed, respectively. At a bare minimum your backend needs to return a JSON response containing a Boolean success property, which influences which callback is triggered. This example expects an additional msg field to let the user know what's going on:

```
{ "success" : true, "msg" : "Thank you for your submission."}
```

Likewise, if your server-side code deems the submission unsuccessful for any reason the server should return a JSON object with the success property set to false. Unlike Ext JS there's no way in Sencha Touch to provide additional feedback or validation information for individual form fields. You'll simply have to roll your own in this case and handle that in your callback.

Handling form validation

A common paradigm is to validate the data in your form before submitting. Doing so ensures that all required fields are filled out and that any specific validations, like an email field containing an email address, are true. In Ext JS achieving this task was a breeze using the `form.validate` function. As of this writing Sencha Touch doesn't have any equivalent function in its arsenal, forcing you to get creative. One way around this is to put validations on the model, like we covered in chapter 7, and then before submitting pull the data out of the form using the `getValues` function and feed it into a new instance of the model. This allows you to use the model validators to validate your form.

Alternatively you can visit the website for this book at www.senchatouchinaction.com, where you can find additional plug-ins and extensions that alleviate this problem as well.

8.5.2 Loading data into your form

The use case for just about every form includes saving and loading data. In the good ol' Ext JS days that was done using the load method, which follows the same paradigm as submitting data with the submit method. Sencha Touch isn't like Ext JS in this regard. It removes the generic load method, replacing it with a setRecord method that consumes a Model instance. This new methodology forces you to use a model and proxy to load data first and then feed it into the form. To illustrate this concept you'll once again visit listing 8.5. Before you can begin you must have data to load, so let's dive right into creating some sample data. Save the following JSON data in a file called data.txt:

```
[
    {
        "firstname"   : "Anthony",
        "lastname"    : "De Moss",
        "description" : "Sencha Touch In Action Launch Party!!",
        "size"        : "medium",
```

```
        "partydate"   : {
            "year"  : 2011,
            "month" : 2,
            "day"   : 23
        }
    }
]
```

Unlike the response for the form submission the `model.load` method doesn't require a success property to be present in the JSON response, but instead expects the data to be wrapped in additional square brackets to create an array of records. To consume the JSON object from earlier you need to first create a model:

```
var PersonModel = Ext.define('Person', {
    extend : 'Ext.data.Model',
    config : {
        fields : [
            'firstname',
            'lastname',
            'description',
            'size',
            'partydate'
        ],
        proxy  : {
            type   : 'ajax',
            url    : 'data.txt',
            reader : 'json'
        }
    }
});
```

You register your model and give it a name and fields that correspond to the names of the form elements from listing 8.5. You define an Ajax proxy on the model itself and point it to the data.txt file that you created earlier. Notice that you assign the return value of `Ext.regModel` to a variable you name `PersonModel`. This approach gives you a few benefits when doing the load. Each registered model exposes a few functions that can be called without needing an instance of `Model`, requiring, though, that you have a handle on the registered model itself. Alternatively, you could've used `Ext.ModelManager.getModel` instead and fed it the name of your model to get a handle on the model itself. Next up is the function that handles the loading:

```
var formLoad = function () {
    PersonModel.load(123, {
        success : function (record) {
            myForm.setRecord(record);
        }
    });
};
```

Here you create a `formLoad` function to take care of loading data remotely via the model you registered. Keep this function somewhat generic so it can easily be called from anywhere in the code. You reference the `PersonModel` you stored earlier and call the `load` method. The `load` method expects two parameters: the unique ID of the record you want to load, and the config object that contains a `success` handler and a `failure` handler, to be triggered based on whether the `load` operation completed successfully.

The `success` handler function receives the loaded record (an instance of your `Model`) as its only parameter. You in turn feed that into the form using the `setRecord` method. Voilà! You have data in your form, just like in figure 8.7.

In the event you already have the data on hand, let's say because you made a separate Ajax call, or use data from another component such as a list, you can set the values of the form fields using `myForm.setValues(dataObj)`, where `dataObj` represents a JSON object equivalent to the content you stored in data.txt, minus the square brackets.

TIP To retrieve the values from any given form, call `getValues` from the `Form Panel` instance. For example, `myForStillm.getValues()` would return an object containing keys representing the names of the fields and their values.

Loading data can be as simple as that.

Congratulations! You've now configured your first truly complex form, learned how to load and save its data, and are familiar with the basics of forms in general. It's

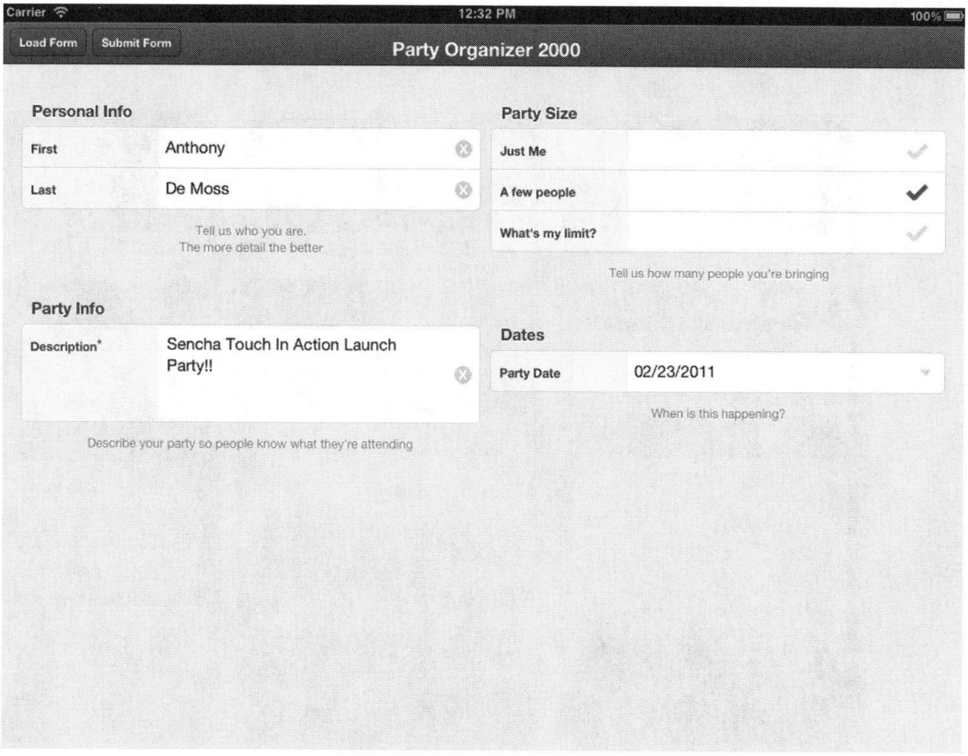

Figure 8.7 Your multicolumn form loaded with data

time you put all this knowledge together and see what you can do when you use forms with other components, like a list.

8.6 Binding a form to a list

Standalone samples and isolated test cases are a great way to illustrate a particular concept, but let's be honest: they never cover the interesting stuff or represent real-world scenarios. Our next sample should remedy that to some degree by utilizing portions of the list sample from chapter 7, panels from chapter 4, as well as code from listing 8.1. The app you're building is a simple party invitation manager (see figure 8.8) that shows a list of invitees, which are bound to a form to allow editing.

Although this scenario is still contrived it represents a starting point that could easily be adapted for other situations. Figure 8.8 illustrates the different screens your app will have, beginning with the list view on the left, to the edit screen in the middle, to the exposed picker for the invitation status on the right. Hitting Save on the form posts the changes back to the list and makes them visible in real time.

The circle you see to the left of the names is used as an indication to display the attendee's RSVP status. Red means that the attendee declined and green is used to indicate that the person has accepted the invitation. Gray indicates that the person is undecided.

To render the circle in your `List` implementation you'll need to bake some custom CSS, as shown in the following listing.

Figure 8.8 The different screens for the party invitation manager

Listing 8.7 The CSS for your custom `List` implementation

```
<style type="text/css">
    .invite-list .status {                          ❶ Adds base style
        display: inline-block;
        vertical-align: middle;
        width: 20px;
        height: 20px;
        border-radius: 10px;
        margin-right: 10px;
    }

    .invite-list .decline {                         ❷ Introduces status colors
        background-color: red;
    }

    .invite-list .undecided {
        background-color: grey;
    }

    .invite-list .accept {
        background-color: green;
    }
</style>
```

The CSS in Listing 8.7 is responsible for rendering the invitation status indicator ❶ and different status colors ❷.

In the next listing you're going to implement your custom CSS by telling the list to use your custom `invite-list` CSS class and use data to drive each item's invitation status.

Listing 8.8 A `List` implementation using your custom CSS

```
Ext.Viewport.add({                                  ❶ Adds list to viewport
    xtype      : 'list',
    fullscreen : true,
    cls        : 'invite-list',                     ❷ Uses custom invite-list CSS class
    itemTpl    : '<span class="status {decision}"></span>{name}',
    items      : {
        xtype  : 'toolbar',
        docked : 'top',
        title  : 'Party Invitation Manager'
    },
    store : {                                       ❸ Creates data store with dummy data
        data : [
            {
                name     : 'Jay Garcia',
                decision : 'decline'
            },
            {
                name     : 'Anthony De Moss',
                decision : 'undecided'
            },
            {
                name     : 'Sebastian Stirling',
```

```
                decision : 'accept'
            }
        ]
    },
    listeners  : {
        itemtap : function (listView, index, target, record) {   <-- On item tap shows sheet with form (4)
            sheet.down('formpanel').setRecord(record);
            Ext.Viewport.add(sheet);
            sheet.show();
        }
    }
});
```

To render your custom list you call the add method ❶ of the Viewport class and pass it an anonymous configuration object that creates an instance of a list view and renders it full screen. You don't need a reference to this List in the rest of your program, so this is the best pattern.

This List instance uses your custom CSS because you configure the cls property ❷, setting it to the base CSS invite-list. Each list item will then use the status CSS because the itemTpl has the token "{decision}", which gets filled in by the data items you configured in your dummy data store ❸.

The last bit to focus on in this listing is the itemtap listener ❹. This function is fired when the itemtap event is fired by the List view instance. The fourth argument is the data record that's used to render each list item. You bind that record to the form that resides in the sheet that you have yet to create. You also add the sheet to the viewport and then tell it to appear.

The next logical step would be to create the sheet, but the configuration to include the toolbar and its Save and Cancel buttons, along with the Form panel and its input fields, requires a sizable amount of code. It's for this reason that you'll be breaking up the sheet creation into three steps.

Because you've already done this stuff these next three listings will be a relatively easy read. First, let's look at the following listing.

Listing 8.9 Defining the toolbar and form for the pop-up sheet

```
var sheetToolbar = {
    xtype  : 'toolbar',                  <-- Configures toolbar XType (1)
    docked : 'top',
    title  : 'Edit Invitee',
    items  : [
        {
            ui      : 'default',
            text    : 'Cancel',
            handler : function (btn) {      <-- (2) Hides sheet
                btn.up('sheet').hide();
            }
        },
        {
            xtype : 'spacer'
        },
```

```
        {
            ui      : 'confirm',
            text    : 'Save',
            handler : function (btn) {
                var form   = sheet.down('formpanel'),
                    record = form.getRecord();

                form.updateRecord(record);
                btn.up('sheet').hide();
            }
        }
    ]
};
```

❸ Persists changes, updates list

❷ Hides sheet

To configure the toolbar create a named reference (`sheetToolbar`) that points to an `XType` config object ❶. This object contains the Cancel and Submit button configurations. The Cancel button has a handler that hides the `Sheet` instance ❷. The Submit button handler is responsible for updating the record with the form data and then hiding the sheet ❸.

With the toolbar configuration complete let's move on to the Form panel configuration in the next listing.

Listing 8.10 The Form panel configuration

```
var sheetForm = {
    xtype    : 'formpanel',
    defaults : { labelWidth : 65 },
    items    : [
        {
            xtype : 'textfield',
            label : 'Name',
            name  : 'name'
        },
        {
            xtype   : 'selectfield',
            label   : 'Status',
            name    : 'decision',
            options : [
                {
                    text  : 'Undecided',
                    value : 'undecided'
                },
                {
                    text  : 'Accepted',
                    value : 'accept'
                },
                {
                    text  : 'Declined',
                    value : 'decline'
                }
            ]
        }
    ]
};
```

❶ Configures Form panel

❷ Includes input fields

❸ Sets up select field

The Form panel configuration ❶ in listing 8.10 includes two input fields ❷, one of which is a select field ❸. This is where attendees will be able to set their invitation status with a few taps.

Now that you have the Form panel configuration complete let's wrap things up with the sheet configuration. You've already done all of the hard work, so this one will be simple:

```
var sheet = Ext.create('Ext.Sheet', {
    layout        : 'fit',
    modal         : true,
    hideOnMaskTap : false,
    height        : 170,
    width         : 310,
    items         : [
        sheetToolbar,
        sheetForm
    ]
});
```

You create the `Sheet` instance right away in this code snippet and implement the toolbar (listing 8.9) and Form panel (listing 8.10) configurations. It's worth noting that the sheet doesn't get displayed until the `itemtap` event is fired from the list you created in listing 8.8.

There you have it! Binding a list to a form requires a bit of typing, but the difficulty level is minimal.

8.7 Summary

In focusing on the `Form Panel` class we've covered quite a few topics, including many of the commonly used fields, ways to arrange forms in creative ways, and most important, how to deal with form data. Although forms in Sencha Touch don't have features like validation checking you can easily add such features by means of extensions or plug-ins. If you need such features check out the Sencha Touch extensions and plug-in forums. Oftentimes community members publish their code, free of charge, for you to use in your applications.

Moving forward, in the next chapter you'll get a look at the exciting world of maps and media before finally topping it all off with a larger application.

9 Maps and media

This chapter covers

- Using the map component
- Using the image component
- Exploring multimedia

So far in this book you've learned how to create and lay out components and retrieve data from a backend to populate data views and lists. Feels pretty good to get this far, doesn't it? Let's push a bit further and see what you can do with Sencha Touch and modern mobile devices. This chapter explores maps and how to use the Google Maps API, the simple `Img` (image) component, and the `Media` class and its subclasses, `Audio` and `Video`. First up is the map component.

9.1 Maps in your application

Google Maps are often used in mobile applications. Have a "Contact Us" section of your application? It could be a smart decision to have a map to show your users the location of your business. In figure 9.1 you see an application that uses Google Maps markers to show house listings in the Houston, Texas area. What you may not know is that the functionality to add markers is surprisingly easy. You may think that it's difficult, but with Sencha Touch's map component and Google Maps and its API it's not that hard. In this section you'll take a look at the Sencha Touch map component, create a simple map, and dive into using the Google Maps API to add markers.

9.1.1 *Maps under the hood*

At first looking at what `Ext.Map` does under the hood can be confusing. You have the Sencha Touch map (`Ext.Map`) component but you also have the Google Map component, making it hard to understand which map component you need to use. To make it worse, `Ext.Map` also uses `Ext.util.Geolocation`, which is a utility class to get GPS coordinates and other useful information via the HTML5 Geolocation spec. So as you can see it may be confusing, but we'll take it slow and discuss each piece separately and then bring them all together at the end.

Figure 9.1 This example shows a map using markers for points of interest. This application was created by the Houston Association of Realtors to show housing listings.

9.1.2 *Location awareness*

When the World Wide Web Consortium (W3C) was working on the HTML5 family one of the highly touted specs was the Geolocation spec. This spec provided rules about how devices should allow the browser to get the current location. Sencha Touch has a utility class called `Ext.util.Geolocation` that takes advantage of this spec and allows you to poll to get location updates.

> **W3C Geolocation**
>
> You can learn more about this spec directly from the W3C's website. It's a technical description but allows you to truly understand what's going on in the background. The URL is http://dev.w3.org/geo/api/spec-source.html.

A good thing to know is that the Geolocation spec does require the browser to ask for permission to get the location. The first time a user tries to get the location the browser will prompt the user for permission to allow the application to receive location updates. This setting will be saved so it won't keep requesting permission for access to location data from the user each time your code attempts to retrieve it. Let's dive into an example of how to use the `Ext.util.Geolocation` class in the next listing.

Listing 9.1 Using the `Geolocation` class

```
var geo = new Ext.util.Geolocation({
    allowHighAccuracy : true,
    listeners         : {
```

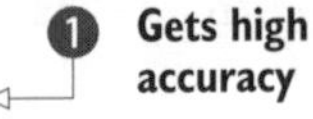

```
        locationupdate : function(geo) {                          ❷ Listens for
            console.log('New latitude: ' + geo.getLatitude());       locationupdate
        },
        locationerror  : function(geo, timeout) {                 ❸ Handles
            if (timeout){                                            error
                console.log('Timeout occurred.');
            } else {
                console.log('Error occurred.');
            }
        }
    }
});                                                               ❹ Starts
geo.updateLocation();                                                polling
```

Listing 9.1 shows how easy it is to use the Sencha Touch `Geolocation` class. By default the location's accuracy may not be that great. It would be good for general location, but if you need more accuracy then you need to set the `allowHighAccuracy` config to `true` ❶. Be aware that this config has adverse effects on the time it takes to get a location and may use more of the device's battery. Getting location is asynchronous, so you have to listen for the `locationupdate` ❷ event to handle when a location is finally retrieved and for the `locationerror` ❸ event because sometimes it may fail. In the `locationupdate` event you can get the latitude and longitude coordinates among other details about the current location. The `locationerror` can fire either because of a failure to connect to the GPS satellites and timing out or just because something went wrong. Just creating the `Geolocation` class won't cause anything to happen, so you have to execute the `updateLocation` ❹ method. You have to execute this method each time you want to get a new location, but by default the `autoUpdate` config is set to `true`, so it will automatically poll and fire the `locationupdate` event. The frequency of this automatic update is based on the `frequency` config, which is 10000 (10,000 ms) by default.

Now that you know how to get your location let's look at how you can create a map.

9.1.3 Creating a simple map

We need to cover just a few quick things before looking at some code. In your index.html file add a `<script>` tag to load the Google Maps API. Without this nothing will work and you'll receive error messages. This is an example of what the addition should look like:

```
<script
  type='text/javascript'
  src='http://maps.google.com/maps/api/js?sensor=true'
></script>
```

The following snippet shows how to create a map component:

```
Ext.create('Ext.Map', {
    useCurrentLocation : true,
    mapOptions         : {
```

```
        mapTypeControl    : false
    }
});
```

That's it! Just six lines of code and you have a map that uses your GPS location. Technically even if you don't have any configuration options set you'd still have a map rendered. But you set `useCurrentLocation` to `true` so that you could use `Ext.util.Geolocation` to determine your current location and center the map on that location. If `useCurrentLocation` is set to `true` the map won't render until the location is found. You then specified the `mapOptions` configuration to an object to configure the Google Maps instance. You set the `mapTypeControl` option here to ensure that the user can't switch between the different map types that would otherwise be available, like a satellite view. You can see more configuration options that can go into this `mapOptions` configuration by visiting http://code.google.com/apis/maps/documentation/javascript/reference.html#MapOptions. Figure 9.2 shows the results.

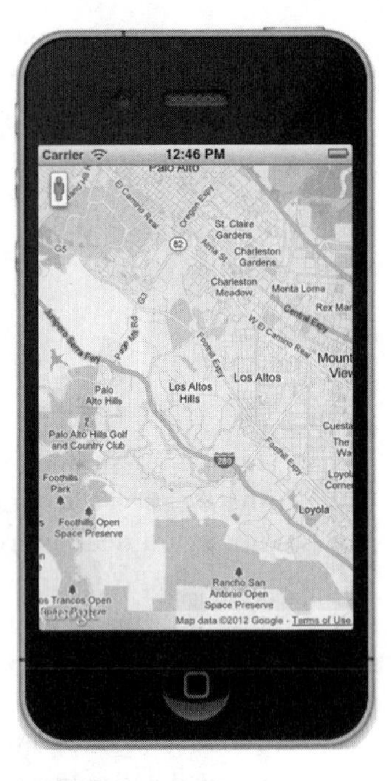

Figure 9.2 A simple map component in action

9.1.4 Getting advanced with Google Maps API

You just learned how to create a simple map. So far, so good, but now let's get a little more advanced and learn how to incorporate more features from the Google Maps API. We won't get too in-depth at this point, so you'll just learn how to add a marker to the map and display an InfoWindow to give some information about what the marker is for. To start let's set the foundation by creating a basic `Ext.Map` instance:

```
var onMapRender = function() {},
    map         = Ext.create('Ext.Map', {
        fullscreen         : true,
        autoUpdate         : false,
        useCurrentLocation : true,
        listeners          : {
            maprender : onMapRender,
            single    : true
        }
    });
```

In this example you don't need automatic updating of your position, so you set `autoUpdate` to `false`. If you don't need automatic updating but left this set to `true` you'll be wasting CPU cycles, memory, and battery by polling the current location every 10,000 milliseconds, and you don't want to waste performance and the user's battery. You want the map to render centered on your current location so you set `useCurrentLocation` to `true`. Finally, you can't add any markers to a map that isn't rendered, so you have to listen for the `maprender` event, and after its first firing it'll remove the listener by using the `single` event option. The `maprender` event is fired on `Ext.Map` when the Google Map is inserted and rendered. Just like you set `autoUpdate` to `false` to avoid wasting resources, you can remove the `maprender` listener. You know you can save some memory because it's only going to fire once.

So you have a map ready, but it's just a plain map, and you want a Google Maps marker to show where on the map you are. The Google Maps marker also gives some context to the map. Without the marker you'd just have a plain map and your users might not know what they're looking at.

You created a simple map and added a listener to the `maprender` event, but the `onMapRender` function is an empty function and you need to take action within that function to show the marker. You want to have a marker render at your current location, and when a user clicks or taps on that marker, have a Google InfoWindow appear to let the user know why the marker is there. See the following listing.

Listing 9.2 Creating a marker and InfoWindow

```
var onMapRender = function (touchMap) {
    var maps   = google.maps,
        map    = touchMap.getMap (),
        geo    = touchMap.getGeo(),                              // 1 Gets Geolocation class instance
        lat    = geo.getLatitude(),
        lng    = geo.getLongitude(),
        coords = new maps.LatLng(lat, lng),
        marker = new maps.Marker({
            position : coords,
            map      : map
        }),                                                      // 2 Creates Google InfoWindow instance
        infowindow = new maps.InfoWindow({
            content : 'This is the current location'
        });

    maps.event.addListener(marker, 'click', function() {       // 3 Adds click event listener to marker
        infowindow.open(map, marker);
    });
};
new Ext.Map({
        fullscreen          : true,
        autoUpdate          : false,
        useCurrentLocation  : true,
        listeners           : {
            maprender   : on MapRender,
            single          : true
        }
});
```

In listing 9.2 you update the `onMapRender` function that gets fired from the `maprender` event. First you cache the `google.maps` namespace onto the `maps` variable for minification and very tiny, minimal performance gain purposes. You then need to get the location, so you use `getGeo` (1) on the `touchMap` argument, which is the `Ext.Map` instance and returns the `Geolocation` instance. To get the latitude and longitude you execute the `getLatitude` and `getLongitude` methods to retrieve those coordinates from the `Geolocation` class. You need to create a Google `LatLng` instance based on the latitude and longitude coordinates so the marker knows where to show in the Google Map. Finally, the marker is created, passing in the `LatLng` instance as the `position` config and the Google Map instance as the `map` config. If you were to stop there you'd have a marker show on the map, but you also want an InfoWindow to describe the marker. So you then create the `InfoWindow` instance ❷, passing in some content that can hold HTML (but be careful of how much you put in there, because the user may be on a cell phone that has limited screen real estate). You then add a `click` listener on the marker with the `addListener` method ❸ so that you can open the InfoWindow using the `open` method, passing in the Google Map instance and the marker. Figure 9.3 shows what listing 9.2 yields.

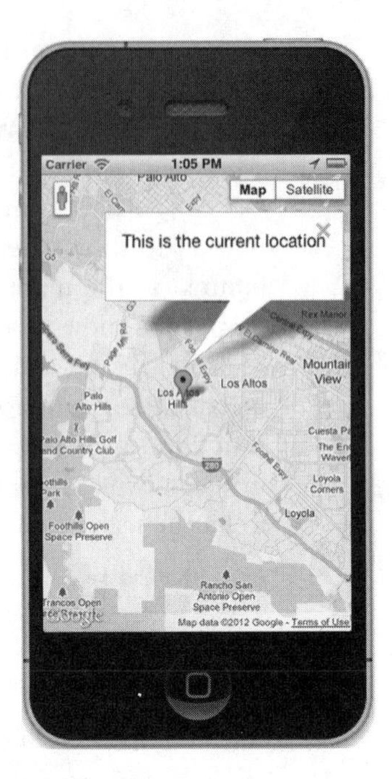

Figure 9.3 Google Maps marker added to the map. Tap on the marker to view the Google InfoWindow.

That's it! You should now have a general understanding of how to incorporate the Google Maps API with the Sencha Touch map component, yet there's so much more you can do with the Google Maps API. You've only used marker and the InfoWindow, but you can show directions between two points or get information about the elevation of a certain coordinate. Reference the Google Maps API documentation to see the different classes you can use. Google even provides some samples. Next up, some fun with the `Img` component!

9.2 Handling images

You may be thinking, "Why is there an image component within Sencha Touch? It's so basic." We agree. Images are such a basic thing that there may not be a need. The reason Sencha Touch has an `Img` component is to provide a way to show an image from a simple configuration and also provide a load and error event to let you know when an image is loaded or when loading has failed. For the `Img` component, getting an image to play nice in layouts would require some extra steps, and Sencha Touch has removed those extra steps and added some features that you can use to create stunning things.

9.2.1 Image basics

The `Img` component simplifies the job of having images be held in a layout, but it also complicates the issue—it's not just a simple component setting an image as the HTML of the component. There are two modes to the `Img` component: background and image. The background mode sets the `background-image` CSS rule to the URL of the image you pass in on the `<div>` element. The image mode will create a `<img>` tag and set the `src` attribute to the URL of the image you pass in; it's wrapped in the `<div>` element. So you can tell the two modes may accomplish the same thing, but they go at it in two different ways. Here's how to define the `Img` component:

```
Ext.create('Ext.Img', {
    width  : 100,
    height : 100,
    src    : 'path/to/image.png'
});
```

By default the `Img` component is in background mode, so if you were to inspect the DOM you'd notice a single `<div>` element with the `background-image` CSS rule set to the `src` config. Now let's look at the image mode and the results (shown in figure 9.4):

```
Ext.create('Ext.Img', {
    width  : 100,
    height : 100,
    src    : 'path/to/image.png',
    mode   : 'image'
});
```

Now if you were to inspect the DOM you'd see an `<img>` element wrapped in a `<div>` element, but if you compare the two end results you may not see a difference. Let's remove the `width` and `height` configs from both instances and check the end result. Notice that the image mode has a size but the background mode `Img` component doesn't. Why is that? It's because the `<img>` tag is telling the `<div>` that it has a size, and due to the styles on the `<div>` it's allowing itself to be autosized based on the size

Figure 9.4 Visually both mode configurations look the same; background mode is on the left and image mode is on the right. Check out the DOM and each `Img` component will look different.

of the <img> element. With the background mode `Img` component the <div> element can't get a size from its CSS `background-image` rule.

More advanced users will also note that you can do different things with each DOM structure. Using the default background mode you can tile the image so it'll repeat horizontally and vertically or just one direction using CSS. Also with CSS you can center the background image, but if you used the image mode you'd have to use more CSS to get the centering effect.

Moving forward, you have other functionality. For example, the `Img` component exposes a load and error event to notify you if the image was loaded or if there was an error, along with a tap event that will fire when it detects it has been tapped on. You can also use the `setSrc` method to change the URL of the image on the fly and even use `setMode` to change between background and image mode on the fly. Another hidden gem of the `Img` component is that when it's hidden the image is removed from the DOM, but because the browser will cache the image when the `Img` component is shown again it'll reset the `src` attribute and the image will appear instantly. This will free up some memory because the image doesn't need to be in memory when hidden, a great feature of Sencha Touch 2.

9.2.2 Preloading an image with a spinner

Showing a static image with the `Img` component is easy. Let's do something fun and use what you've learned so far in this book. You'll build a proof of concept example in listing 9.3, where you display a preloading spinner while the image you want to show loads. You'll create a container with two `Img` components. The first item of the container will be the preloading spinner `Img` component that has been Base64-encoded so that it doesn't have to be downloaded but can be shown right away. The second item will be the image you want to show, but it may take a second or more to download on a mobile cellular network that usually has high latency. The Base64-encoded spinner image won't be shown here because it's quite a long string; this image comes with Sencha Touch and is located at /sencha-touch-2.0.0/resources/themes/images/default/loading.gif.

Listing 9.3 Image preloading spinner example

```
Ext.create('Ext.Container', {
    fullscreen : true,
    items      : [
        {
            xtype  : 'image',                          ❶ Adds preloading image
            height : 250,
            width  : 250,
            cls    : 'loading-image',
            src    : 'data:image/gif;base64,-------'
        },
        {
            xtype     : 'image',                           ❷ Adds real image
            height    : 250,
            width     : 250,
```

```
                hidden     : true,
                src        : 'path/to/image.png',
                listeners : {
                    delay  : 2000,
                    load   : function(image) {
                        var container = image.up('container'),
                            spinner   = container.down('image');

                        container.remove(spinner);                    ❸ Removes preloading image
                        image.show();                          ❹ Shows real Image
                    }
                }
            }
        ]
});
```

In listing 9.3 you use a container to wrap the two `Img` components. You have two image items under the container, and once the image is loaded you'll remove the loading image and show the image. The first item is the `Img` component ❶ for the preloading spinner image that uses the Base64-encoded string of the loading.gif animated GIF image that comes with Sencha Touch 2. The second item is the image ❷ that you want to show, but because you want control over when that `Img` component is shown you have it marked as hidden. You use a URL to load the image from. You set the height and width of both `Img` components; you did this so that the size will remain the same when you switch to the `Img` component in the `load` event listener.

Because you're only creating a proof of concept example right now you're using a computer, and that means you have a reliable internet connection. So to see if it all works you can use the `delay` event option to delay the firing of the `load` event listener by 2,000 milliseconds. Within the `load` event listener you get the `Img` component instance as the first argument, so you can resolve the container by using the `ComponentQuery` convenience method `up` on the `Img` component instance. Now that you have the `Container` instance you can get the preloading spinner `Img` component by using the `ComponentQuery` convenience method `down` on the container. Because the preloading spinner `Img` component is the first child item of the Container and the `down` method will return the first instance it finds you know it'll return the correct `Img` component. The `load` event fired on the actual `Img` component, so you don't need to preload the spinner `Img` component. That means you can go ahead and remove ❸ the preloading spinner `Img` component so that the memory it's using can be used for other things. Finally, you show the `Img` component ❹.

Run the code replacing the `src` configs with the Base64 string and a URL to a remote image, and possibly update the width and height configs. For a remote image URL we found a nice Sencha logo at www.sencha.com/files/blog/old/blog/wp-content/uploads/2010/06/sencha-logo.png, which is 250px by 250px. Works pretty well, huh? Figure 9.5 shows the before and after image loading from this test.

This preloading spinner is a great candidate to be made into its own class to handle this internally. To do this you'd use `Ext.define` to define a new class, which we'll cover in chapter 10.

Figure 9.5 The loading spinner (left) and the actual image loaded (right)

9.3 Mastering media

Another feature of an application that could be a large selling point is adding multimedia: audio and video. Maybe you have an application that helps teach a user a subject. Reading text will work, but playing audio or (even better) playing a video to walk the user through something is a great tool to have, and your users would most likely keep coming back because of this multimedia experience. In this section we'll explore both the audio and video components, but first we need to cover the base class, `Media`.

9.3.1 Media base

As you learned back in chapter 3, Sencha Touch has a Component model where components extend other components to introduce common functionality among subclasses (call abstraction) and dependability. The audio and video components both extend from the media component. So what does the media component give the audio and video components?

First you must understand what's going on under the hood. HTML5 has both `<audio>` and `<video>` DOM elements, and there are many common features of these tags, like common events and ways to control them. The media component is a wrapper of these so that its subclasses can have this functionality without duplicating the code. Now let's take a look at the common functionality in the media component.

Events are usually a large part of your app, so having some useful events is critical. Think of a DVD player; it plays, pauses, stops, rewinds, and fast-forwards, basic functionality that every model must have. The media component bubbles `play`, `pause`, `ended`, and `timeupdate` events from the underlying `<audio>` or `<video>` so you have

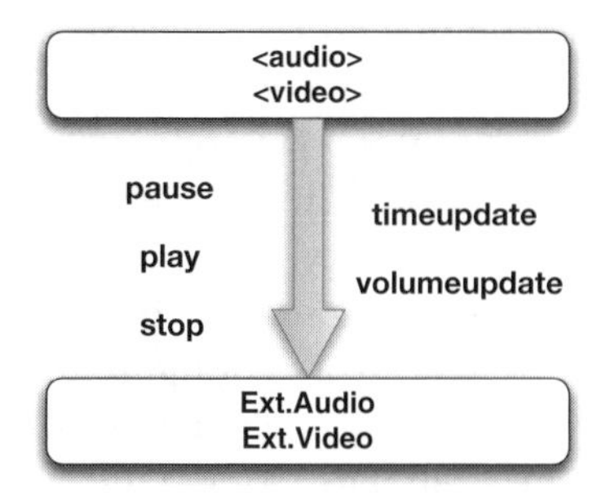

Figure 9.6 The five events that get bubbled from the `<audio>` and `<video>` DOM elements to the `Ext.Audio` and `Ext.Video` Sencha Touch components

events for the basic functionality that a media component should have, as shown in figure 9.6. You also have a `volumechange` event that'll fire when you use the HTML5 controls to change the volume. Along with the `volumechange` event you get a `mutedchange` event that'll fire when the `<audio>` or `<video>` element is muted or unmuted. There are quite a few useful events to build multimedia functionality into your application.

NOTE The `volumechange` event won't fire if you change your device's volume. This event will only fire if you use the onscreen control to change the volume of the media or use the `setVolume` method.

Table 9.1 shows some of the most commonly used configuration settings.

Table 9.1 Common configurations

Config	Description	Defaults
`url`	The URL of the media file to be used	Empty string, `''`
`enableControls`	`true` to show the onscreen controls to play/pause, seek, and change the volume	Android – `false` Others – `true`
`autoResume`	`true` to start playback when the component has been visually activated	`false`
`autoPause`	`true` to stop playback when the component has been visually deactivated	`true`
`preload`	`true` to begin the preloading of the media immediately	`true`
`loop`	`true` to loop the media forever	`false`
`volume`	The volume level of the media accepting a value between 0 (no volume) and 1 (full volume)	`1`
`muted`	`true` to mute the media volume	`false`

The most frequently used config will be the `url` config so that you can play a media file. If you don't want controls on screen so that you can change the volume or seek through the media file you can set `enableControls` to `false` (`enableControls` defaults to `true` unless you're on Android, in which case it defaults to `false`). If you want the media to start playing when the media component is visually activated you can set autoResume to true; it defaults to false. Users may not want to wait around for media and may decide to jump to another section of your application without stopping the playing of the media. The autoPause config, which defaults to true, will pause the media when the media component has been visually deactivated, saving a lot of resources on the device. You also have a loop config that'll loop the media when the media reaches the end; it defaults to true. To configure the volume you have a muted config that can start the media muted (it defaults to false), and you also have a volume config that accepts a value from 0 to 1 (it defaults to 1). You can visit Sencha Touch 2's API documentation for these configuration settings and more. Setting up the media component and listening for its events is just two-thirds of the functionality. Let's look at some of the methods to control the media component programmatically.

Table 9.2 shows some of the methods associated with the most commonly used configuration settings.

Table 9.2 The associated get and set methods of the configurations from table 9.1

Config	Get Method	Set Method	Returns/Accepts
url	getUrl	setUrl	String
enableControls	getEnableControls	setEnableControls	Boolean
autoResume	getAutoResume	setAutoResume	Boolean
autoPause	getAutoPause	setAutoPause	Boolean
preload	getPreload	setPreload	Boolean
loop	getLoop	setLoop	Boolean
volume	getVolume	setVolume	Int (0-1)
muted	getMuted	setMuted	Boolean

Each of the configurations has a set method associated, with the first letter of the config uppercased and set as the prefix. For example, you can change the URL on the fly with the setUrl method. If the media was playing at the time the setUrl method fired it'd start the playing of the new media. You also get the setEnableControls, setAutoResume, setAutoPause, setLoop, setMuted, and setVolume methods from the configs. With these alone you can see how much control you now have over the media, and that's not all the control you have. You also have play, pause, and stop methods, as well as a toggle method. The toggle method toggles the playback status between playing and paused. Another method to control the media is the setCurrentTime method, which allows you to jump around the timeline of the media.

Not all methods are used to control the media; you also have methods to get information about the media. These methods are getDuration, getCurrentTime, and isPlaying. getDuration will tell you the length of the media in seconds. getCurrentTime will tell you where in the media the current time is in seconds. To find out if the media is currently playing you can use the isPlaying method, which will return a Boolean whether or not the media is playing. Like the set methods from the configs, you also have the get methods from the configs: getUrl, getEnableControls, getAutoResume, getAutoPause, getLoop, getMuted, and getVolume.

Wow, that's a lot of functionality! Remember, this is just the base class of the audio and video components, so each of them has all of these configs, events, and methods, providing a lot of control over your media. We just covered Media; now let's look at its first subclass, Audio.

9.3.2 Listening to audio

Because the media component has all that functionality the audio component doesn't have any more functions to provide. The audio component is basically just a configuration component to set up the DOM elements that the functionality from the media component can control. The following listing shows how to set up the audio component shown in figure 9.7.

Listing 9.4 Basic audio example

```
Ext.create('Ext.Audio', {
    fullscreen : true,
    url        : 'assets/audio/audio.mp3',
    autoResume : true,
    loop       : true,
    volume     : 0.5
});
```

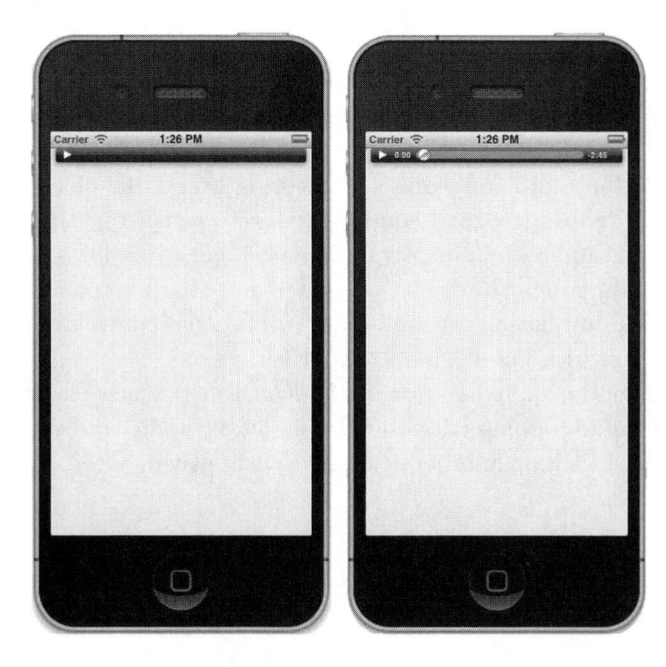

Figure 9.7 The audio views before playback (left) and after pressing the play button on the onscreen controls (right)

Listing 9.4 shows how easy it is to set up an audio component. You give a URL to an MP3 file to play. When the audio component is visually activated you want to set `autoResume` to `true`. You don't want the audio to stop playing, so set `loop` to `true` so that it'll just continue to play instead of stopping at the end. You never know how loud the user has his volume set so setting the `volume` configuration to half volume or even less may be a good idea, as the user could change this at any time.

To be a little more advanced you could put the audio component within a container and give the container a docked toolbar with some buttons to control the playback and volume, as shown in figure 9.8, using what you know about the control methods that the media component brings. Let's look at some of these functions.

Figure 9.8 An example of an audio component in a container with some toolbars to handle simple playback and volume change

There's one thing to note about using the audio component. It can only work if the device, or more specifically the operating system, can play the audio file. Not every audio file and codec is supported on a mobile operating system. For instance, iOS and Android don't support the Windows Media formats like WMA. The recommended formats for cross-device compatibility are uncompressed WAV and AIF, MP3, AAC-LC, and HE-AAC, with MP3 probably the most common format supported. If the audio component isn't playing more than likely the codec isn't supported, and trying one of the other recommended formats could serve as a quick test to debug the issue. You'd be amazed at how many times someone has complained that the audio component wasn't working when it turned out that the device's volume was turned down or muted.

Because this is only for audio you won't see anything except the onscreen controls unless you set the `enableControls` config to `false`. Even if the MP3 file has album artwork the HTML5 audio won't display this. You can get around this by using a container with an `Img` component to display an image and an audio component as children of the container. Just having the audio with its onscreen controls may look a bit underwhelming to your user, but this is up to your UI.

Those are the basics of the audio component. Sencha Touch makes playing audio simple. Hopefully your mind is going crazy thinking of all the audio stuff you can do in your application. Let's kick it up another gear and look at playing video.

9.3.3 *Playing video*

Like the audio component the video component extends the media component, so `Video` gets all the same methods, configurations, and events. The video component adds one extra feature: the use of a poster. A *poster* is an initial image that's used instead of showing the video; tapping on the poster will start playing the video. This is a departure from the `<video>` tag specification, but the reason Sencha Touch does this is due to device resource limitation. It takes quite a lot of resources to handle video; the device has to decode the video file and then play it. Scrolling around an application would feel choppy and wouldn't provide your users with a great experience. To get around this Sencha Touch shows a simple image (poster) in place of the video, allowing for application performance to be much better. Of course, once that video is playing the benefits of the poster will be lost. Let's take a look at creating a `Video` instance in the next listing.

Listing 9.5 Basic video example

```
Ext.create('Ext.Video', {
    fullscreen : true,
    url        : 'assets/audio/video.mp4',
    posterUrl  : 'assets/poster/video.png',
    autoResume : true,
    loop       : true,
    volume     : 0.5
});
```

You first give the video the URL of an MP4 video file and also use the URL of a PNG image to serve as the poster. You use `autoResume` to resume video playback on activation and `loop` to keep the video playing endlessly. You set `volume` to 0.5 so that you don't play video with overly loud sound. Figure 9.9 shows the video component on an iPhone.

Figure 9.9 The video component on an iPhone with the poster (left) and in the native video player when playing the video (right)

Figure 9.10 The video component running on an iPad, where you can see the poster (left) and the video playing inline in the application (right). (Source: Blender Institute, part of the Blender Foundation)

Once again, not every mobile operating system can play every video file format and codec. What does all this mean to you as a web developer? It means there's no standard video format in the HTML5 video spec. Both iOS 5.1 and Android 4 currently support an MP4 file with H.264 as of this writing, but Google has pulled H.264 support in its popular Chrome browser for personal computers, so the future of H.264 support in Android is unknown. As of this writing Google has made no announcement that it'll pull H.264 support in Android. We'd be surprised if iOS stops H.264 support any time soon, even if a competing codec is adopted in the HTML5 video spec. Figure 9.10 shows the video component playing with a tablet.

HTML5 video history

At first Ogg Theora was recommended as the video codec standard for HTML5, but some unknown patents have scared large companies like Apple away from supporting it. Interestingly, the H.264 codec may also have unknown patents, but H.264 has been widely used so the perceived threat to mobile operating system makers has decreased to almost nothing. Even though H.264/MPEG-4 AVC has been widely used it hasn't been made part of the HTML5 video specification because users have to pay licensing fees to the MPEG LA group. Microsoft and Apple are among the members of this group.

iOS is made by Apple, an MPEG LA group member, so Apple doesn't have to worry about the licensing fees as much as Google (creator of the Android operating system) does. Google isn't part of the MPEG LA group but it still supports the H.264 codec. Google acquired On2, and thus the VP8 codec, and provided a royalty-free license to users of the VP8 codec. Google combined the VP8 video codec with the Vorbis audio format within a Matroska container, creating the WebM format. Microsoft and others have criticized the WebM format, but Mozilla and Opera have asked for the inclusion of WebM into the HTML5 video specification.

9.3.4 Things to keep in mind

We need to go over a few things about using media in your application. Keep in mind that you're on a mobile network that has high latency and isn't as reliable as home broadband. Having several audio or video components load at the same time will choke the connection on your user's device. You have no way to know what else users will be doing or what their network connection will be like. Your users could be on a network that isn't great and their connection may be slow, so loading audio and video files could take forever. So loading one at a time and optimizing files for mobile devices is recommended. For instance, a video file that looks great on a high-definition, 1080p television is way too much for a mobile device. You can keep 1080p high-definition, but the video size should be smaller because tablets and phones don't have large screens and it'd be a waste to have a large-sized video.

Performance is another issue. Today's devices are more powerful than ever, with more memory, quad-core CPUs and GPUs, and mobile connections that rival home broadband connections. But even these mobile devices can still be strained with large video files and you can't always be certain that your users will have the latest and greatest devices. A good application will keep performance in mind with every feature implemented and be tested on older devices. We know that people with iPhone 3 devices view our applications so we still need to support these devices. As long as you put performance first you'll have no problem finding ways to have every feature you want.

Using poster images for videos will help performance quite a bit. Scrolling with inline videos without a poster image will provide a bad experience, making your application look like it has performance issues. The user doesn't know that the device is having problems keeping everything rendered at once. Using a poster image, the device doesn't have to try to keep videos rendered and keep up with the scrolling; images are a tremendously smaller strain on the device.

One more thing to know is that each device is going to handle videos differently. The iPad will play the video inline with your application (depicted in figure 9.10 earlier) as you'd want, but the iPhone with the same iOS version will launch the video in the native video player (depicted in figure 9.9) so the video isn't actually playing within your application. This isn't going to affect you and your application, because when users press the Done button they'll return to your application where they left off. Nevertheless, it's something to be aware of, because this is the behavior of the device rather than your application and it's what your users will experience. We suggest trying to view your application on a variety of devices so you know the experience your users will have and you can make changes based on your results and what you want your users to experience.

9.4 Summary

This chapter added a considerable amount of fun to your applications. You first looked at how the map component works and you learned how to incorporate the

Google Maps API. The pair can be used together to give your application a great user experience. You then learned about the `Img` component and saw that it's much more than meets the eye. You created a proof of concept using a preloading spinner image that displays while the real image loads. Finally, you explored `Media` and its subclasses, `Audio` and `Video`. You learned that the `Media` class has a lot of functionality that the audio and video components inherit. You saw how to create and set up `Audio` and `Video` instances and also what codecs can be used. Learning about the history of the HTML5 video spec allows you to be intelligent while developing your application, and you learned that playing videos can have consequences for the device and its mobile connection. By using what you learned in this chapter you can improve the user experience of your application by adding map and media functionality.

In the next chapter we'll take a closer look at how to use basic JavaScript prototypal inheritance, and then you'll learn to master the Sencha Touch class system. You'll also create a plug-in while you're at it.

Part 3

Constructing an application

Part 3 is aimed at getting your Sencha Touch knowledge up to the level of professionals, beginning with chapter 10.

In chapter 10 you'll tackle advanced concepts like prototypal inheritance with JavaScript and the Sencha Touch class system. You'll also have the opportunity to learn how the dependency injection system works. You'll take that knowledge and create a custom list widget as well as a custom plug-in.

Chapter 11 is focused on creating applications. Here you'll learn how to use Sencha Cmd to manage an application from start to finish. We'll also walk you through the development of an application where you'll get to experience complex theories like developing with the Sencha Touch application architecture and MVC.

By the end of part 3 you'll have learned many of the secrets to implementing advanced features in the Sencha Touch framework.

10 Class system foundations

This chapter covers

- Understanding prototypal inheritance
- Developing your first extension
- Knowing how plug-ins work
- Developing a real-world plug-in
- Looking into the Sencha Touch class loader

Every Sencha Touch developer faces challenges where reusability is an issue. Often, a component of an application is required to appear more than once within the application's usage lifetime. Without mastering these techniques you could end up with what's known as "function soup," or unmaintainable code. This is why we'll focus on the concept of reusability by the use of framework extensions and plug-ins.

In the first section of this chapter you'll learn the basics of extending (subclassing) with Sencha Touch. You'll begin by learning how to create subclasses with JavaScript, where you'll see what it takes to get the job done with the native language tools. This section will give you the foundation to refactor your newly created subclass to use the Sencha Touch class system.

Once you're familiar with creating basic subclasses you'll focus your attention on extending Sencha Touch components. You'll have fun learning the basics of framework extensions, and you'll solve a real-world problem by extending the grid Panel widget and see it in action.

When you finish the extension you'll learn how extensions solve problems but can create inheritance issues when similar functionality is desired across multiple widgets. After you've gotten your hands dirty with the creation of a custom extension you'll dive into creating a plug-in. In this exercise you'll learn how to inject the ability for fieldsets to be collapsible in any of your applications.

After you've built a solid foundation in the Sencha Touch class system you'll take a few moments to look at the dynamic class loader that Sencha Touch provides. You'll learn about the three popular patterns for using the dynamic class loader and the caveats that go along with each one.

10.1 Classic JavaScript inheritance

JavaScript provides all the necessary tools for prototypal inheritance, but it falls short when it comes to easily setting up multiple class inheritance. Sencha Touch makes multiple class inheritance much easier with the class system. To begin learning about inheritance, you'll create a base class.

To help you along, envision that you're working for an automobile dealership that sells two types of cars. One type is the base car, which serves as a foundation to construct the premium model. Instead of using 3D models to describe the two car models, you'll use JavaScript classes.

NOTE If you're new to object-oriented JavaScript or are feeling a bit rusty, the Mozilla foundation has an excellent article to bring you up to speed or polish your skills. You can find it at https://developer.mozilla.org/en-US/docs/JavaScript/Introduction_to_Object-Oriented_JavaScript.

You'll begin by constructing a class to describe the base car, as shown in the next listing.

Listing 10.1 Constructing your base class

```
var BaseCar = function(config) {                  // 1 Creates constructor
    this.octaneRequired = 86;
    this.shiftTo = function(gear) {
        this.gear = gear;
    };
    this.shiftTo('park');
};
BaseCar.prototype = {                              // 2 Assigns prototype object
    engine     : 'I4',
    turbo      : false,
    wheels     : 'basic',
    getEngine  : function() {
        return this.engine;
    },
```

```
    drive      : function() {
        return 'Vrrrroooooooom - I'm driving!';
    }
};
```

In listing 10.1 you create the `BaseCar` class constructor ❶ which, when instantiated, sets the instance's local `this.octaneRequired` property, adds a `this.shiftTo` method, and calls it, setting the local `this.gear` property to `'park'`. Next, you configure the `BaseCar`'s `prototype` object ❷, which contains three properties that describe the `BaseCar` and two methods.

You could use the following code to instantiate an instance of `BaseCar` and inspect its contents with the WebKit JavaScript console:

```
var mySlowCar = new BaseCar();
mySlowCar.drive();
console.log(mySlowCar.getEngine());
console.log('mySlowCar instance:', mySlowCar);
```

Figure 10.1 shows what the output of this code looks like in the WebKit JavaScript multiline editor and console.

```
Vrrrroooooooom - I'm driving!
I4 mySlowCar contents:
▼ BaseCar
    gear: "park"
    octaneRequired: 86
  ▶ shiftTo: function (gear) {
  ▼ __proto__: Object
    ▶ drive: function () {
      engine: "I4"
    ▶ getEngine: function () {
      turbo: false
      wheels: "basic"
    ▶ __proto__: Object
```

Figure 10.1 Instantiating an instance of `BaseCar` and exercising two of its methods

With your `BaseCar` class set you can now focus on subclassing it. First you'll do it the traditional way. This approach will give you a better understanding of what's going on under the hood when you use `Ext.define` later.

10.1.1 Inheritance with JavaScript

Creating a subclass using native JavaScript is achievable with multiple steps. Rather than simply describing them we'll walk through the steps together. The following listing creates `PremiumCar`, a subclass of the `BaseCar` class.

Listing 10.2 Creating a subclass the old-school way

```
var PremiumCar = function() {
    PremiumCar.superclass.constructor.call(this);      ❶ Calls superclass constructor
    this.octaneRequired = 93;
};

PremiumCar.prototype   = new BaseCar();                ❷ Sets subclass prototype
PremiumCar.superclass  = BaseCar.prototype;            ❸ Sets the superclass

PremiumCar.prototype.turbo  = true;
PremiumCar.prototype.wheels = 'premium';
PremiumCar.prototype.drive = function() {
   this.shiftTo('drive');
   PremiumCar.superclass.drive.call(this);
};
PremiumCar.prototype.getEngine = function() {
    return 'Turbo ' + this.engine;
};
```

To create a subclass you begin by creating a new constructor which is assigned to the reference PremiumCar. Within this constructor is a call to the constructor method of PremiumCar.superclass within the scope of the instance of PremiumCar being created (this).

You do this because, unlike other object-oriented languages, JavaScript subclasses don't natively call their superclass constructor ❶. Calling the superclass constructor gives it a chance to execute and perform any constructor-specific functions that the subclass might need. In our case, the shiftTo method is being added and called in the BaseCar constructor. Not calling the superclass constructor would mean that your subclass wouldn't get the benefits provided by the base class constructor.

Next, you set the prototype of PremiumCar to the result of a new instance of BaseCar ❷. Performing this step allows PremiumCar.prototype to inherit all of the properties and methods of BaseCar. This is known as *inheritance through prototyping*, and it's the most common and robust method of creating class hierarchies in JavaScript.

In the next line you set the PremiumCar's superclass reference to the prototype of the BaseCar class ❸. You can then use this superclass reference to do things like create so-called extension methods, such as PremiumCar.prototype.drive. This method is known as an extension method because it *calls* the like-named method from the superclass prototype, but from the scope of the instance of the subclass it's attached to.

TIP All JavaScript functions (JavaScript 1.3 and later) have two methods that force the scope execution: call and apply. To learn more about them, visit www.webreference.com/js/column26/apply.html.

With the subclass now created you can test things out by instantiating an instance of PremiumCar with the following code entered into the WebKit JavaScript console editor:

```
var myFastCar = new PremiumCar();
myFastCar.drive();
console.log('myFastCar contents:');
console.dir(myFastCar);
```

Figure 10.2 shows what the output would look like in the WebKit JavaScript multiline editor and console.

This output shows that your subclass performed as desired. From the console.dir output you can see that the subclass constructor set the octaneRequired property to 93, and the drive extension method even set the gear method as "drive".

This exercise shows that you're responsible for all of the crucial steps in order to achieve prototypal inheritance with native JavaScript. First you had to create the constructor of the subclass. Then you had to set the prototype of the subclass to a new instance

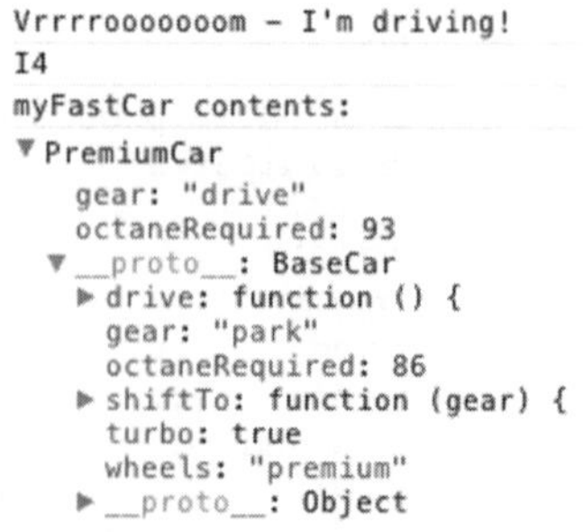

```
Vrrrroooooooom - I'm driving!
I4
myFastCar contents:
▼ PremiumCar
    gear: "drive"
    octaneRequired: 93
  ▼ __proto__: BaseCar
    ▶ drive: function () {
      gear: "park"
      octaneRequired: 86
    ▶ shiftTo: function (gear) {
      turbo: true
      wheels: "premium"
    ▶ __proto__: Object
```

Figure 10.2 Your PremiumCar subclass in action

of the base class. Next, for convenience, you set the subclass's `superclass` reference. Last, you added members to the prototype one by one.

You can see that quite a few steps need to be followed in order to create multiple classes with the native language constructs. We think this is a lot of work and it frustrates us, so luckily there's an easier way to achieve this result.

Next you'll see how the Sencha Touch class system makes creating classes and multiple inheritance much easier.

10.2 Using the Sencha Touch class system

The Sencha Touch class system takes JavaScript's prototypal inheritance to another level, adding features like dependency injection, automatic setter and getter method creation, statics, and mix-in support (multiple inheritance). All of these features require the use of Sencha Touch class-specific methods, such as `Ext.define`, `Ext.create`, and `Ext.require`, as you'll learn later. As we walk through this section you're going to learn why the Sencha Touch class system is a great solution for developing your applications.

10.2.1 Using Ext.define

You just saw what it takes to implement JavaScript prototypal inheritance. You had to do quite a bit of work just to get a single level of inheritance set up. With complex software like your applications you'd have to do quite a bit of typing to get inheritance going. This means that you'd have a lot of redundant code in your projects, resulting in a lot of bloat.

Sencha Touch is in the perfect position to take on the heavy lifting for you, and it even adds a bit of automation. To see what we mean you'll have to take a step back to the first two classes you created. Begin by redefining the `BaseCar` class you constructed in section 10.1. Along the way we'll introduce you to the "config system." Afterward you'll extend the `BaseCar` class to define `PremiumCar`.

> **You'll be using namespaces!**
> Instead of just using basic class names you'll be defining your class names using properly namespaced class names, much like you find in classic languages such as Java or C++. If you're new to namespaces, they help organize your code much like folders do in a file system. When developing applications, you'll need to organize your code according to namespace, so it's good to get used to it now.

The next listing shows the `BaseCar` class defined in the `MyApp.car` namespace. You'll extend `Ext.Component` because that's where you'll spend a lot of your time when developing applications.

Listing 10.3 Defining your base class

```
Ext.define('MyApp.car.BaseCar', {
    extend       : 'Ext.Component',
    requires     : 'Ext.Component',
```

(1) Defines BaseCar class

```
    config      : {
        octaneRequired : null,
        gear           : null,
        engine         : 'I4',
        turbo          : false,
        wheels         : 'basic'
    },
    constructor : function() {
        this.callParent();
        this.setOctaneRequired(86);
        this.setGear('park');
    },
    drive       : function() {
        console.log("Vrrrroooooom - I'm driving!");
    }
});
```

❷ Defines config object

❸ Sets the constructor

Listing 10.3 demonstrates the most basic use of `Ext.define`, where you define `MyApp.car.BaseCar` ❶. The first thing that might strike you as strange is the way you define the class names: by string. You do so because Sencha Touch gives you an opportunity to define a class and namespace together. Sencha Touch will create the `MyApp.car` namespace for you if it hasn't been previously defined, and it'll place the `BaseCar` class within that namespace. This pattern paves the way for using the class loader system later on, where you'll learn how important it is to organize your classes in your project's file system according to the namespace for which they're defined. For now, let's finish looking at this class. We still have a lot to discuss.

In the class definition you instruct the class system to extend `Ext.Component`, which is now known as the superclass for your `BaseCar` class.

> **Why extend Ext.Component?**
>
> You're extending `Ext.Component` because it gives you the capabilities to use the automatically generated setters and getters. Most of your work will be done extending classes that have these magic functions created, so it makes sense to start getting used to it now.

You also instruct Sencha Touch to require `Ext.Component`. Doing so allows you to fully utilize the class loader dependency injection system which is absolutely necessary to use the Sencha SDK, which we'll cover in chapter 11.

Next you define what's known as the class `config` object ❷. This object is somewhat special and is used to define custom properties that'll be set and accessed after instantiation. When you define classes, setter and getter methods are automatically generated for you. The `config` object also gives you the capability to define custom properties in a single place, instead of in the prototype of the class.

In the constructor ❸ you immediately call the superclass's constructor. Doing so allows the superclass to do what it needs to do in its own constructor.

You can see this in action by instantiating an instance of your custom class. To instantiate this class you'll have to use `Ext.create` instead of the JavaScript `new` keyword. By now you should be used to using `Ext.create` for Sencha Touch classes, but we want you to use it within the context of your own class. Here's how you do it:

```
var mySlowCar = Ext.create('MyApp.car.BaseCar');
mySlowCar.setGear('drive')
mySlowCar.drive();
console.log(mySlowCar.getEngine());
```

Figure 10.3 shows the results of implementing the `MyApp.car.BaseCar` class, which are exactly as you'd expect.

```
Vrrrroooooooom - I'm driving!
I4
```

Figure 10.3 The results of our first Sencha Touch class

Notice how you're using the `setGear()` and `getEngine()` methods. These are magic methods automatically generated by the class system. Right away you can start to see how the Sencha Touch class system brings power to your code by generating API methods for your custom class, and there's more to the class system than what we just covered.

With your custom class instantiated, you can extend it using `Ext.define`. In doing so you'll see more of how the class system works. The following listing contains the code for the `PremiumCar` class, which extends your `BaseCar` class.

Listing 10.4 Extending `BaseCar` with `Ext.define`

```
Ext.define('MyApp.car.PremiumCar', {          ❶ Defines PremiumCar class
    extend      : 'MyApp.car.BaseCar',
    config      : {                            ❷ Overrides BaseCar config primitives
        turbo  : true,
        wheels : 'premium',
        stereo : '5.1'
    },
    constructor : function() {                 ❸ Creates PremiumCar constructor
        this.callParent(arguments);
        this.setOctaneRequired(93);
    },
    applyEngine    : function(engine) {        ❹ Adds apply method
        return 'Turbo ' + engine;
    },
    drive       : function() {                 ❺ Extends drive method
        this.setGear('drive');
        this.callParent();
        console.log('The turbo makes a big difference!');
    }
});
```

In listing 10.4 you create the `MyApp.car.PremiumCar` class, which is an extension (subclass) of the `MyApp.car.BaseCar` superclass, using the `Ext.define` method. Here's how this works.

You first call on `Ext.define` and define the `MyApp.car.PremiumCar` ❶ extension class. You instruct Sencha Touch to extend your previously defined `MyApp.car.BaseCar` class by naming it via string representation set to the `extend` keyword.

Next you set the config object ❷ to override the turbo, wheels, and stereo properties inherited from the superclass. Even though you override those three properties the other config properties from the superclass will be inherited.

Managing inheritance for config properties

One of the things we want you to recognize and remember is that the class system will manage what's known as "deep merging of config properties." That is, the extension class will inherit properties that are defined in the superclass. This is very much like basic prototypal inheritance except it adds those automatically generated setters and getters for you.

When thinking about extending classes you must consider whether prototypal methods in the subclass will share the same name as prototypal methods in the base class. If they'll share the same symbolic reference name you must consider whether they'll be extension methods or overrides.

An extension method is a method in a subclass that shares the same reference name as another method in a base class. What makes this an extension method is the fact that it includes the execution of the base class method within itself. The reason you'd want to extend a method would be to reduce code duplication, reusing the code in the base class method.

The constructor ❸ for this class is an extension method. It's an exact duplicate of the previously created PremiumCar constructor, with the addition of the call to the parent class constructor via this.callParent(arguments);. This statement allows the subclass to chain the constructor method calls, effectively allowing the MyApp.car.BaseCar superclass constructor to execute within the scope of new instances of your MyApp .car.PremiumCar subclass.

You're doing something a bit different in this class, adding what's known as an apply method ❹ or applier function. An apply method is called by a setter method and gives your class an opportunity to do something with the value being set. In our example you're prepending Turbo to the set engine type. So every time the autogenerated setEngine() method is called the engine type will always have turbo applied.

A unique class inheritance pattern

Internally Sencha Touch uses the config system for every class, and apply methods are in a lot of the classes. For the Component base class apply methods are used to do things like construct instances of classes via a factory method known as Ext.Factory. The reason apply methods exist is to abstract the construction of instances of classes from a setter method.

The drive method ❺ is an extension method as well. In this method you automatically set the gear to drive and then call the parent method. Doing so allows you to easily call `drive` on instances of this extension class and it automatically sets the gear for you.

> **Overriding methods**
> Even though you're not using it it's good to discuss what an override is. An *override* is a method in a subclass that shares the same name as another method in a superclass but doesn't chain method calls up to the superclass via `this.callParent()`. You override a method if you wish to completely discard the code that's in the like-named method in the base class. Therefore you don't call `this.callParent()` within the method and the upwards execution chain won't occur.

Now that you have your `PremiumCar` configured using `Ext.define` you can see it in action using the WebKit JavaScript console. You can do so using the same code you used when exercising your manually created subclass:

```
var myFastCar = Ext.create('MyApp.car.PremiumCar');
myFastCar.drive();
console.log(myFastCar.getEngine());
```

Figure 10.4 shows what it looks like in the WebKit JavaScript console.

```
Vrrrroooooom - I'm driving!
The turbo makes a big difference!
Turbo I4
```

Figure 10.4 The results of the instantiation of the `PremiumCar` class

You've just successfully extended a class using `Ext.define`. What we've shown you are just the basics of using `Ext.define`, where you created a class from scratch (`MyApp.car.BaseCar`) and extended it (`MyApp.car.PremiumCar`).

From here you can extend the `MyApp.car.PremiumCar` class and create a `MyApp.car.SportsCar` class that adds features like `drift()`-ing and perhaps `dragRace()`-ing.

What you've done thus far with prototypal inheritance with JavaScript and `Ext.define` positions you to start exercising your newly found knowledge and start extending Sencha Touch.

10.3 Extending Sencha Touch components

Extensions to the framework are developed to introduce additional functionality to existing classes in the name of reusability. The concept of reusability drives the framework, and when utilized properly it can enhance the development of your applications.

Some developers create preconfigured classes, which are constructed mainly as a means of reducing the amount of application-level code by stuffing configuration parameters into the class definition itself. Having such extensions relieves the application-level code from having to manage much of the configuration, requiring only the simple instantiation of such classes. This design pattern is okay but should only be used if you're expecting to stamp out more than one instance of this class. Otherwise it's considered wasteful to define a class just as a home for a collection of configuration parameters.

Other extensions embed behavioral logic inside the class itself or add features such as utility methods. An example of this would be a Form panel that automatically pops up a message box whenever a save operation failure occurs. We often create extensions for applications for this very reason, where the widget contains limited built-in behavioral logic. We say "limited" because with Sencha Touch 2.0 you now have an MVC architecture for which you can abstract business logic to controllers. This is something you'll learn about in the next chapter.

We're going to develop an extension to Sencha Touch's `List` widget and really dive into some complex concepts such as configuring and controlling CSS3 animation and callbacks.

Instead of attempting to recreate the example code in the rest of this chapter, I'd like you to follow along in the example code that you got from the zip file as you read the listings here. The code for this section can be found in examples/chapter10 and for clarity is located in the following files:

- examples/chapter10/actionslist.html
- examples/chapter10/actionlist.css
- examples/chapter10/js/ActionList.js
- examples/chapter10/js/ActionListItem.js
- examples/chapter10/icons/icons.css

Let's move on, shall we? This is going to be fun!

10.3.1 Thinking about what you're building

Many phone and tablet applications use lists of some sort. Often to invoke an action (such as delete or update) you tap on a list item, and another screen appears allowing you to invoke that action. It turns out that there's a better way to invoke actions, and it all has to do with a swipe gesture on a list item. With a bit of work you can extend the `List` and `ListItem` classes to allow you to inject this little bit of functionality, making your applications much more intuitive and easier to use. To make the picture clearer figure 10.5 illustrates your extension sliding the actions for a list item into view.

As figure 10.5 depicts, swiping on a row will cause the row actions to appear via slide animation. Tapping on any action icon that's not the dismiss (X) icon will cause an event to be fired by your custom list implementation, allowing your app do things like call someone. Tapping on the dismiss icon will cause the action row to close. The action row will also be dismissed if the user decides to scroll the list or selects or swipes another row.

The Sencha Touch List widget renders instances of `ListItem` to display data from the Store. This means that in order to achieve our goal we'll need to extend `ListItem` and `List`. Our extension to `ListItem` will be known as `ActionListItem` and will be responsible for managing the horizontal swipe gesture. Our extension to `List` will be named `ActionList` and will implement our `ActionListItem` and work with it to manage things like disabling of the vertical scroll bar, since horizontal swipe and vertical scroll gestures conflict somewhat.

Figure 10.5 Four frames demonstrating our first extension in action, where we flip-slide an action into view for a list item via a swipe event and tap on an action to display the Message Box alert dialog

You'll code this extension in four phases:

- Create the CSS and discuss the action icons
- Build the `ActionListItem` extension
- Generate the `ActionList` extension
- Implement your `ActionList`

Let's begin coding, shall we?

10.3.2 Getting the CSS and icons out of the way

To get your extension to work you'll have to generate some CSS to render the icons in their proper place. To do this you'll have to create a CSS style sheet.

Create a CSS file and add the contents of the following listing to it. Don't forget to include your newly created file in your example page.

Listing 10.5 The CSS for the actions

```
.flexbox {
    display             : -webkit-box;
    height              : 47px;
    background-color    : rgb(205, 231, 194);
    -webkit-box-shadow  : inset 0 5px 5px #888;
}

.list-icon {
    margin               : -5px 5px 0;
    background-position  : center center;
    background-repeat    : no-repeat;
    width                : 32px;
    height               : 32px;
```

❶ Orders icons in horizontal row

❷ Styles each icon

```
    margin-top          : 7px;
}

.target-dismiss {
    position : absolute;
    right    : 10px;
}

.opaque-list-item {
    background-color : rgba(247, 247, 247, 1);
    z-index          : 99;
    position         : absolute;
    top              : 0;
}
```

❸ Styles last (remove) icon

❹ Makes the ListItem opaque

The CSS styling for listing 10.5 is relatively simple. It contains CSS styles for the container for the icons ❶ so that they're arranged side to side using the WebKit horizontal box layout. The listing also contains styles for all action icons ❷, allowing them to be properly spaced from each other. We also set up the proper styles for the dismiss icon ❸, ensuring that it's flush to the right. The final style ❹ will be used to prevent the icons from displaying through the `ListItem` element, which is transparent by default. So when it's time to animate the `ListItem` element out of view we'll apply this CSS class to it, ensuring that we get the look that we're trying to achieve.

The last bit of styling we need to discuss before you can start writing JavaScript revolves around icons. To render actions you need a set of icons. We found a free set called Glyphicons (http://glyphicons.com/). The entire set contains 420 icons. To complete this example, copy examples/chapter10/icons from this book's examples zip file that you downloaded and include the icons.css file. Otherwise you won't see actions, and that's no good!

With the styling and icon discussion out the way you can start carving out some JavaScript.

10.3.3 *Creating the ActionListItem class*

Our custom `ActionList` extension requires some support for a class that we'll call `ActionListItem`. This class will be responsible for the horizontal gesture management, animating of the visible row data, and rendering the actions in each row (hence the "Item" suffix in the name), and will be responsible for most of the work.

This class is nearly 200 lines long and is much too large to print and digest in one fell swoop, so we'll look at the first third in one shot, as shown in listing 10.6. This will allow you to get a sense of the basic structure of the class. Afterwards we'll fill in four of the method stubs that I've outlined.

It's worth noting that this is a foundational class and we won't be able to render anything on screen until we implement our `ActionList` class in section 10.3.4.

The full contents of the listings in this subsection can be found in js/ActionListItem.js. Later on when we define our `ActionList` extension we'll require the `ActionListItem` with the class loader dependency injection system.

Listing 10.6 The `ActionListItem` extension with method stubs

```
Ext.define('Ext.ux.ActionListItem', {                    <-- 1 Defines our class
    extend : 'Ext.dataview.component.ListItem',
    xtype  : 'actionlistitem',

    config : {
        actionsEl : null,
                                                          2 Configures the
        actionsTpl : [                                 <-- actions Template
            '<div class="flexbox">',
                '<tpl for=".">',
                    '<div class="list-action list-icon {.}" action="{.}">'
                    '</div>',
                '</tpl>',
                '<span class="target-dismiss list-icon remove_2"' +
                        'action="dismiss"> </span>',
            '</div>'
        ].join('')
    },

    applyActionsTpl : function(cfg) {},                 <-- Creates the actions
    initialize : function() {},                           3 Template instance
    onElementTap: function(eventObj) {},
    onDrag : function(event) {},
    showActions : function(actions, moveToX) {},
    onDragEnd  : function() {},
    hideActions : function() {},
    clearActionsEl : function() {}
});
```

We kick things off with listing 10.6 by defining our custom `Ext.ux.ActionListItem` class ❶. For our class we must extend the `Ext.dataview.component.ListItem` class. To make implementation in the `ActionListItem` easy we set up a custom `XType`.

The config block contains three key items. The first is `actionsEl`, a null property that we're setting to allow us to have automatically generated getter and setter methods. We'll use this to track the element that will contain the actions we'll render with the template defined as `actionsTpl` ❷.

`actionsTpl` initially is an HTML fragment that is converted to an instance of `XTemplate` via the applier function named `applyActionsTpl` ❸. We use the applier functions that are called automatically when the class is defined and when the setter methods are called. Appliers in this context are treated as factories that return instances of something.

In this case, our applier is returning an instance of `XTemplate` for us. The reason we are using an applier here is because at the time our class JavaScript file is parsed by the browser, `Ext.XTemplate` is not available for instantiation. However, we know for a fact that when our `ActionListItem` instances are created, `XTemplate` will be available because it's a core UI class that's used by much of the framework.

When we cover `showActions` I'll show you how we'll use the automatically instantiated `XTemplate` instance to render actions and inject it into the DOM for our `ActionListItem` extension.

The rest of this listing contains stubbed-out methods. Table 10.1 provides a quick rundown of what these methods will do.

Table 10.1 Stubbed-out methods of our custom `ActionListItem` class

Method	Purpose
`initialize`	This method will help bootstrap this class, hooking up necessary element-level event listeners for the `tap`, `drag` and `dragend` events.
`onElementTap`	Whenever the `ActionListItem` element is tapped this method is called. If an action was tapped it will fire a custom event and pass necessary data like the action name and the item's data record.
`onDrag`	This method is called when the component's element `drag` event is fired. It's going to be responsible for doing some math to figure out if the `drag` is horizontal in order to slide out the rendered data and reveal the actions underneath. It will also fire a custom `horizontaldrag` event that will be listened to by the parent `ActionList` class. Given that a single `ActionListItem` is not aware of any siblings, the parent `ActionList` will need to determine if it's OK for an `ActionListItem` to show its actions or not, since it's aware of all child `ActionListItem` instances.
`showActions`	The parent `ActionList` extension will call this method when it's absolutely sure that an `ActionListItem` instance should reveal its actions underneath. This method will be responsible for configuring the CSS3 animation during the `drag` event cycle.
`onDragEnd`	After an `ActionListItem` has engaged in a horizontal drag gesture, the `onDrag` method will set some class-level members, storing critical data to finalize the gesture. If someone swipes the `ActionListItem` left or right they are gesturing to show the actions for that row. Eventually they'll stop dragging the `ActionListItem` element that displays the row data, and we'll need to animate that element to allow the actions to become visible. This method will also be responsible for some cleanup of class-level member properties set by `onDrag`.
`hideActions`	This method will be called by the parent `ActionList` and will animate the data element back into position and, after the animation has completed, will execute `clearActionsEl`.
`clearActionsEl`	Here we'll manage the destruction of the element that is displaying the actions.

Now that we have a basic understanding of what each method is responsible for, let's look at the contents of the `initialize` and `onItemHorizontalDrag` methods in the next listing.

Listing 10.7 The last four methods of our `ActionListItem` extension

```
initialize : function() {
    var me        = this,
        myElement = me.element;
```

```
        me.callParent();

        myElement.on({                                    <-- Registers event
            scope : me,                                       listeners for
            tap   : 'onElementTap'                        ❷ the element
        });

        myElement.on({
            scope      : me,
            drag       : 'onDrag',
            dragend    : 'onDragEnd',
            delegate   : '.x-list-item-inner'
        });
    },                                                    ❸ Handles the element
    onElementTap : function(eventObj) {                <--   tap event
        var me     = this,
            action = eventObj.target.getAttribute('action');

        if (action) {
            if (action != 'dismiss') {
                me.fireEvent('actiontap', me, action, me.getRecord());
            }
            me.hideActions();
        }

        if (me.getActionsEl()) {
            eventObj.stopEvent();
        }
    },
```

The `initialize` method ❶ is responsible for bootstrapping the `ActionListItem` instance. After setting some variable references we chain the superclass `initialize` method via the execution of `this.callParent`. We call the parent `initialize` method so soon in order to allow the parent's `initialize` method to do whatever work it needs to do right away. After the parent `initialize` method execution, we register event handlers on the `ActionListItem` instance's element ❷. The first event handler is a generic `tap` handler, which will call our `onElementTap` method. The next two events are responsible for the dragging of the list item data-representation element, hence the delegate CSS selector property.

`onElementTap` ❸ is responsible for firing an `actiontap` event handler if an action was tapped. It knows this because we've set an action parameter directly on the element via the `actionsTpl` as defined in listing 10.6. If an action was tapped we fire the custom event and call `hideActions` to animate the data-representing element over the actions. This element stops the element `tap` event from bubbling up if we're displaying the actions to prevent the `List` view from attempting to select that row.

Next we'll look at the `onDrag` method, which is shown in the following listing.

Listing 10.8 The `onDrag` method

```
onDrag : function(event) {
    var me     = this,
        currX  = event.getXY()[0],
```

```
            prevX  = me.prevX || currX,
            deltaX = currX - prevX,
            initX  = me.initialDragX,
            moveToX;

        me.prevX = currX;

        // Horizontal Dragging
        if (Math.abs(deltaX) >= 5) {
            me.isDraggingHorizontal = true;

            if (! initX) {
                me.initialDragX = currX;
            }

            if (currX < initX) {
                me.dragDirection = 'left';
                moveToX = currX - initX;
            }
            else if (currX > initX) {
                me.dragDirection = 'right';
                moveToX = Math.abs(currX - initX);
            }
            me.fireEvent('horizontaldrag', me, moveToX);

        }
    },
```

Recall that `onDrag` is going to be executed when the `drag` event is fired from the `ActionListItem`'s element. The `drag` event will be fired for an `ActionListItem` instance while the user is scrolling, so it's essential to figure out whether or not the user is doing vertical dragging (scrolling the data) or horizontal, which indicates that the user wants to reveal the actions underneath.

This method might seem quite complex at first glance, but it's actually very simple. If the `drag` event fires and we're tracking a gesture (think of the `prevX` property), we need to figure out if it's a horizontal swipe gesture. The `drag` event needs to fire at least twice in order for the gesture tracking to work, which is why we cache some local instance properties. We're not using the `config` system here because we want to be as fast as possible, and using the auto-generated getter and setter methods could slow things down as they would add unnecessary call stack overhead.

If the gesture is horizontal and the delta is at least greater than or equal to five pixels, we do some calculations to figure out the direction of the gesture and fire a custom `horizontaldrag` event, passing in the `ActionListItem` instance and X coordinate to move to. This `horizontaldrag` event will be listened to by the `ActionList` class to halt vertical scrolling and hide scroll bars if they are present, as well as execute the `ActionListItem` instance's `showActions` method, thereby completing the animation sequence.

Before we move on, it's important to note the `dragDirection` property being set on the instance of the `ActionListItem` via this method. The `dragDirection` property is set by this method to acknowledge that a horizontal drag event was detected and in which direction. This `dragDirection` property will be used by `onDragEnd` to figure out in which direction to complete the slide-out animation of the element representing the data in the model.

In the next listing we'll cover the showActions method.

Listing 10.9 The showActions method

```
showActions : function(actions, moveToX) {
    var me                = this,
        itemDockBody      = me.element.down('.x-dock-body'),
        itemTextElCt      = itemDockBody.down('.x-list-item-inner'),
        itemTextElCtStyle = itemTextElCt.dom.style,
        itemActionsEl     = me.getActionsEl(),
        actionsTpl        = me.getActionsTpl();

    itemTextElCt.addCls('opaque-list-item');

    if (! itemActionsEl) {
        itemActionsEl = actionsTpl.append(itemDockBody, actions);
        itemActionsEl = Ext.get(itemActionsEl);
        me.setActionsEl(itemActionsEl);

        itemTextElCtStyle.webkitTransitionTimingFunction = 'ease-out';
        itemTextElCtStyle.webkitTransitionProperty = 'translate3d';
        itemTextElCtStyle.webkitTransitionDuration = '.05s';
    }

    itemActionsEl.setWidth(itemTextElCt.getWidth());

    itemTextElCtStyle.webkitTransform = 'translate3d(' + moveToX
             +  'px, 0, 0)';
},
```

showActions is going to be executed by the parent List class in response to the ListItem's horizontaldrag event. At this point, you may wonder why we didn't call showActions from onDrag. The answer is that our application will be interfacing with ActionList, not the ActionList child ActionListItem instances. Therefore, we'll configure the ActionList with the actions to be rendered on the ActionListItem instances, and the perfect time to make the ActionListItem instances aware of the actions to be displayed is at the time when the ActionListItem needs to display them.

This showActions method has two main responsibilities: injecting the DOM for the action icons and animating the data element left or right based on the moveToX parameter passed. It accomplishes these tasks by first setting the appropriate variable references and then adding the 'opaque-list-item' CSS class to the element that renders the data from our model instances. The purpose of this is to ensure that when we do inject the DOM for the actions, the icons do not bleed through the text, causing some really ugly effects.

After the class is applied to the appropriate element, we use the actions XTemplate instance, which was constructed in our applyActionsTpl applier method (listing 10.6), to merge the array of actions—passed as the first argument from our ActionsList instance with the template—and inject it in the ActionListItem DOM. The DOM

```
▼<div class="x-container x-list-item x-stretched x-list-item-relative" id="ext-actionlistitem-2"
style="width: 100% !important; min-height: 47px !important;">
  ▼<div class="x-dock x-dock-horizontal x-stretched" id="ext-element-30">
    ▼<div class="x-dock-body" id="ext-element-31">
      ▼<div class="x-inner x-list-item-inner opaque-list-item" id="ext-element-29" style="-webkit-
      transition: 0.2s ease-out; transition: 0.2s ease-out; -webkit-transform: translate3d(1505px,
      0, 0);">
        ▼<div class="x-unsized x-list-item-body" id="ext-component-6">
            <div class="x-innerhtml " id="ext-element-106">Acevedo, Stone</div>
          </div>
        </div>
      ▼<div class="flexbox" id="ext-element-128" style="width: 1505px !important;">
          <div class="list-action list-icon phone" action="phone"> </div>
          <div class="list-action list-icon thumbs_up" action="thumbs_up"> </div>
          <div class="list-action list-icon thumbs_down" action="thumbs_down"> </div>
          <span class="target-dismiss list-icon remove_2" action="dismiss"> </span>
        </div>
      </div>
      <div class="x-unsized x-item-hidden x-list-disclosure x-dock-item x-docked-right" id="ext-
      component-7" style="display: none !important;"></div>
    </div>
  </div>
```

Figure 10.6 The DOM for our `ActionListItem` class

fragment is injected as a sibling to the element that wraps the DOM representing the data in our model instances. Figure 10.6 shows what the injected DOM looks like inside of the `ActionListItem` instance.

The last bit of work this method accomplishes is the setting of the appropriate CSS3 transition properties to allow the data element to slide left or right based on the `moveToX` parameter.

The next method we'll cover is `onDragEnd` (shown in the next listing), which is responsible for the end-cycle of the `drag` event loop.

Listing 10.10 The `onDragEnd` method

```
onDragEnd   : function() {
    if (! this.dragDirection) {
        return;
    }

    var me                = this,
        itemDockBody      = me.element.down('.x-dock-body'),
        dockBodyWidth     = itemDockBody.getWidth(),
        itemTextElCt      = itemDockBody.down('.x-list-item-inner'),
        itemTextElCtDom = itemTextElCt.dom,
        itemTextElStyle = itemTextElCtDom.style,
        moveToX           = dockBodyWidth;

    if (me.isDraggingHorizontal) {
        me.fireEvent('horizontaldragend', me);
    }

    if (me.dragDirection === 'left') {
        moveToX *= -1;
    }

    itemTextElStyle.webkitTransitionDuration = '.20s';
    itemTextElStyle.webkitTransform = 'translate3d('
         + moveToX  +  'px, 0, 0)';
```

```
    delete me.prevX;
    delete me.prevY;
    delete me.isDraggingHorizontal;
    delete me.initialDragX;
    delete me.dragDirection;
},
```

onDragEnd is charged with managing the end of the drag event loop and its sole responsibility is to complete the animation of the element that represents the data from the ActionListItem model. It seems complex at first glance, but it's rather simple. The first thing we do is check to see if this.dragDirection is set on the instance of ActionListItem. If it's not, we simply abort the execution of this method. Recall that drag events occur for ActionListItem for both the scrolling of the data and the horizontal gesture we're trying to implement. dragDirection is set by the onDrag method but only if a horizontal drag gesture is present. As you may already see, the dragDirection property is essential for figuring out how to set up the remaining CSS3 slide-out animation.

After the initial if-condition, we set up a bunch of lexically scoped references to elements in the DOM that represent the ActionListItem as well as the dimensions of the element known as the "dock body." We need these so that we can finish off the slide-out animation of the data-representing element, which is done by adding the webkitTransitionDuration and webkitTransform properties directly on the data-representing element's CSS object.

The final bit of work that's accomplished is removal of members that were used to track the drag gesture, set by the onDrag method (listing 10.9), via the JavaScript delete keyword operator.

This wraps up the bit of code that's responsible for revealing the actions. The next step is to worry about hiding the actions. This is where we'll end the ActionListItem class, looking at the hideActions and clearActionsEl methods in the next listing.

Listing 10.11 The hideActions and clearActionsEl methods

```
hideActions : function() {
    if (! this.getActionsEl()) {
        return;
    }

    var me                  = this,
        itemDockBody        = me.element.down('.x-dock-body'),
        itemTextElCt        = itemDockBody.down('.x-list-item-inner'),
        itemTextElCtDom     = itemTextElCt.dom,
        itemTextElCtStyle   = itemTextElCtDom.style,
        transitionListener = function() {

            itemTextElCtDom.removeEventListener(
                  'webkitTransitionEnd',
                   transitionListener
            );
            itemTextElCtStyle.webkitTransitionDuration = '0s';
```

```
                itemTextElCt.removeCls('opaque-list-item');
                me.clearActionsEl();
            };

        itemTextElCtDom.addEventListener(
            'webkitTransitionEnd',
            transitionListener
        );

        itemTextElCtStyle.webkitTransform = 'translate3d(0,0,0)';
    },

    clearActionsEl : function() {
        Ext.destroy(this.getActionsEl());
        this.setActionsEl(null) ;
    }
```

Finally, we land on the `hideActions`. This method is simply responsible for animating the first card back into view, hiding the actions that we rendered and displayed via the `showActions` method. It does this by means of setting up references, very much like `onDragEnd` did, with the addition of a transition callback function, known here as `transitionListener`.

We create the transition callback in this way because we want to be able to perform some post-animation cleanup, such as deregistering the listener after the animation is done, as well as a few other things like removing the '`opaque-list-item`' class from the data-representing element. It's important to note that `clearActionsEl` is called from this callback handler.

After the references are set, we apply the `webkitTransitionEnd` listener via the typical DOM `addEventListener` method, and then tell the data-representing element to translate back to its home position via the setting of the CSS style object's `webkitTransform` property.

`clearActionsEl` is responsible for the destruction of the element that renders the actions, as well as setting the local actions member to null via a call to `this.setActions`. We do this so we can perform immediate cleanup of the DOM and release resources.

This wraps up the `ActionListItem` foundational class. In order to see it working we need to write `ActionList`, which will implement `ActionListItem`.

10.3.4 Creating the ActionList class

We've just defined a major dependency, `ActionListItem`. As you'll soon see, defining the `ActionList` is going to be rather easy, since `ActionListItem` does the bulk of the work to make this extension work.

The role of `ActionList` is not only to implement instances of `ActionListItem` but to perform tasks like managing the enabling and disabling of the scroll when the horizontal drag is occurring on an `ActionListItem`, as well as propagating the `actiontap` event from the `ActionListItem` instances.

This class is 86 lines, which is much too long to fit on one page, so we'll break it up much like we did with the `ActionListItem` class that we just covered. The following file will be created as js/ActionList.js.

Listing 10.12 The `ActionList` class

```
Ext.define('Ext.ux.ActionList', {                         <-- ❶ Defines ActionList
    extend   : 'Ext.dataview.List',
    xtype    : 'actionlist',

    requires : [ 'Ext.ux.ActionListItem' ],               <-- ❷ Requires ActionListItem

    config : {
        defaultType         : 'actionlistitem',           <-- ❸ Implements ActionListItem
        actions             : null,
        currentActionsItem  : null,

        control : {                                       <-- ❹ Registers event handlers on ActionListItem
            'container > actionlistitem' : {
                actiontap          : 'onActionItemTap',

                // Custom events fired by ActionListItem
                horizontaldrag     : 'onItemHorizontalDrag',
                horizontaldragend  : 'onItemHorizontalDragEnd'
            }
        }
    },

    initialize : function() {                             <-- ❺ Initializes ActionList
        var me = this;
        me.callParent();

        me.getScrollable().getScroller().on({
            scope  : me,
            scroll : 'onScrollClearActionsEl'
        });

        me.on({
            scope  : me,
            select : 'onItemSelectClearActionsEl'
        });
    },

    onItemHorizontalDrag : function(item, moveToX) {},
    onItemHorizontalDragEnd : function() {},
    onScrollClearActionsEl : function() {},
    onItemSelectClearActionsEl  : function() {},
    onActionItemTap : function(listItem, action, record) {}
});
```

The listing above contains quite a bit in a very small package. It begins with the definition of our custom `Ext.ux.ActionList` class ❶, which extends `Ext.dataview.List`. We set the `XType` to 'actionlist' and we'll use it for lazy instantiation in just a little bit.

We inject the `ActionListItem` dependency by setting it in the `requires` array ❷. This ensures that the `ActionListItem` class is loaded and stays defined in the namespace before we get a chance to instantiate it.

Focusing on the `config` object ❸, we find that we set our extension's `defaultType` property to the value of 'actionlistitem.' Recall that this is the `XType` of our `ActionListItem` class and `defaultType` is a Container-specific config option. This config object contains two null values. The first is actions, which we'll use for implementation of this class. As we'll see later on, this will be passed an array of strings. The second is `currentActionsItem`, which will be used to track the current `ActionListItem` instance displaying its actions.

The last bit to focus on for the `config` object is the control construct ④. Here we're instructing our class to listen to the `actiontap`, `horizontaldrag`, and `horiztonal-dragend` events of any `ActionListItem` instance present as a child of our `List`. Whenever any of those events are fired, their respective event listener method will be called within the scope of our custom `ActionList` instance. We'll look deeper into these methods later on.

The `initialize` method ⑤ contains a few key things, such as the registration of listeners on the `Scroller` instance, as well as the actual `ActionList` component instance. We listen for the `scroll` event on the `Scroller` instance so we can attempt to clear any visible actions when the user scrolls the list vertically. We listen to the `select` event of this `ActionList` instance so that we can do the same if someone selects an item while another item may be displaying its actions.

The rest of the methods for this `ActionList` class are relatively simple and we'll look at them in the next listing. This listing is somewhat lengthy, but I've formatted it so it's a really easy read.

Listing 10.13 The last five methods for the `ActionList` class

```
onItemHorizontalDrag : function(item, moveToX) {
    var me          = this,
        scrollable  = me.getScrollable(),
        scroller    = scrollable.getScroller(),
        currActions = me.getCurrentActionsItem();

    scroller.setDisabled(true);
    scrollable.hideIndicators();

    item.showActions(this.getActions(), moveToX);

    if (currActions != item) {
        me.onScrollClearActionsEl();
    }
    me.deselectAll();
    me.setCurrentActionsItem(item);
},

onItemHorizontalDragEnd : function() {
    var scrollable = this.getScrollable(),
        scroller   = scrollable.getScroller();

    scroller.setDisabled(false);
},

onScrollClearActionsEl : function() {
    var me          = this,
        actionItem  = me.getCurrentActionsItem();

    if (actionItem) {
        actionItem.hideActions();
```

① Manages the horizontaldrag event

② Handles the horizontaldragend event

③ Clears the actions when scroll occurs

```
            me.setCurrentActionsItem(null);
        }
    },

    onItemSelectClearActionsEl  : function() {      ◁ ❹ Hides the actions when an item is selected
        this.onScrollClearActionsEl();
    },

    onActionItemTap : function(listItem, action, record) {      ◁ ❺ Clears the actions when an item is tapped
        var me = this;

        me.setCurrentActionsItem(null);
        me.deselectAll();

        me.fireEvent('actiontap', me, listItem, action, record);
    }
```

`onItemHoriztonalDrag` ❶ will be executed whenever an `ActionListItem` instance fires its `horizontaldrag` event and will mainly be responsible for disabling the scroller and hiding its UI, as well as hiding any visible actions and clearing any selected items.

When an `ActionListItem` ends its horizontal gesture cycle, it fires a `horiztonal-dragend` event. `onItemHorizontalDragend` ❷ is called from that event and is in charge of re-enabling the scroller for the list.

`onScrollClearActionsEl` ❸ is called whenever the user makes a scroll gesture and is responsible for hiding any visible actions. Similarly, whenever a user selects an item in the list, `onItemSelectClearActionsEl` ❹ is called to hide any visible actions.

Lastly, when a set of actions is visible, if a user taps on an action, the `ActionList-Item` fires an event. When that event is fired, `onActionTap` ❺ is executed and fires an event passing along the data for the tapped `ActionListitem`.

This puppy is now complete and ready to be implemented. This is where we'll get to see two extensions work together with custom CSS!

Listing 10.14 Implementing our `ActionList` class

```
Ext.Loader.setConfig({
    paths : {
        'Ext.ux' : 'js/'
    }
});

Ext.require([
    'Ext.MessageBox',
    'Ext.data.Store',
    'Ext.ux.ActionList'
]);

Ext.define('ListModel', {
    extend : 'Ext.data.Model',
    config : {
        fields : [
            'firstName',
            'lastName'
        ]
```

```
    }
});

Ext.application({
    launch : function() {

        Ext.create('Ext.ux.ActionList', {
            fullscreen : true,
            itemTpl    : '{lastName}, {firstName}',
            items      : {
                xtype  : 'toolbar',
                docked : 'top',
                title  : 'ActionList extension!'
            },
            store      : {
                model    : 'ListModel',
                autoLoad : true,
                proxy    : {
                    type : 'ajax',
                    url  : 'complexData.json'
                }
            },
            actions    : [
                'phone',
                'thumbs_up',
                'thumbs_down'
            ],
            listeners  : {
                actiontap : function(list, listItem, action, record) {
                    var msg = record.data.firstName
                            + ' ' + record.data.lastName;

                    Ext.Msg.alert(
                        'You tapped ' + action,
                        msg
                    )
                }
            }
        });
    }
});
```

❶ Creates instance of ActionList

❷ Configures custom action icons

❸ Listens for the actiontap event

In implementing our custom `ActionList` class ❶ we can configure an array of strings, known as `actions` ❷, that represent action icons to be rendered. To get feedback that an `action` was tapped we simply set up a listener for the custom `actiontap` event ❸. That displays a MessageBox alert dialog showing which record was tapped.

Figure 10.7 shows what it looks like on a phone.

We just saw the power of creating custom extensions with Sencha Touch. With some work we were able to extend two classes to give our applications a capability that's found in many native applications.

The key to making custom extensions like this lies in reading the Sencha Touch documentation and source code. When I first developed this extension I had to

Figure 10.7 Four frames demonstrating our first extension in action, in which we flip-slide an action into view for a List item via a swipe event and tap on an action to display the MessageBox alert dialog

read through the `Ext.dataview.List` source, which led me to look through how it works and uses the `ListItem` class. From there, I stepped through the code and inspected the DOM. With some experimentation I was able to publish this extension.

So, the lesson is, don't ever be afraid to look at the source code. Even if you don't get it at first glance, you'll learn something along the way and will be able to create some pretty awesome UI goodness.

You have your extension out of the way and now you can move on to the last thing you'll do in this chapter: creating a plug-in.

10.4 Creating a Sencha Touch plug-in

You're going to create a plug-in that will allow form fieldsets to be collapsible via a tap of the associated fieldset title. Before you dive into creating plug-ins let's have a quick chat about how plug-ins work.

> **Component life cycle refresh**
> In case you don't remember when exactly plug-ins are created and initialized, now would be an excellent time to brush up on the initialization phase of the component life cycle, which we covered in chapter 3.

10.4.1 The anatomy of a plug-in

The basic anatomy of a plug-in is simple: it starts out by defining your class. If you want to extend an existing Sencha Touch class, it's okay to do so. For your plug-in you'll extend `Component`, because you'll need to inject DOM into your list:

```
Ext.define('MyApp.plugins.MyPlugin', {
  extend : 'Ext.Component,
  alias : 'plugin.myplugin',
  // class related stuff
});
```

The previous snippet demonstrates the basics of creating a plug-in using `Ext.define`, in which you set up `alias` as `'plugins.myplugin'`. You prefix your plug-in alias with `'plugins'` to allow the Sencha Touch class management system to route the registration of this class with `PluginManager`, which will be responsible for creating instances of your classes via lazy objects. These are much like `XTypes`, but they're known as plug-in types in this case.

Here's an example of how you'd use a lazy object to configure this plug-in in a generic `Component` instance:

```
Ext.create('Ext.Component', {
    plugins : [
        {
            type : 'myplugin'
        }
    ]
});
```

In this code you create an instance of `Ext.Component` and set its `plugins` property to an array with a single object. The single object has a `type` property set as `'myplugin'`. When the component nears the end of its initialization phase it'll create an instance of your custom plug-in via this `type` shortcut.

This pattern is considered the best practice by many in the industry, and it uses the Sencha Touch lazy instantiation mechanism. There's an alternative pattern that uses the fully qualified class name, and it looks like this:

```
Ext.create('Ext.Component', {
    plugins : [
        {
            xclass : 'MyApp.plugin.MyPlugin'
        }
    ]
});
```

The difference between the two patterns is that the first one (using `type`) requires the class itself to be defined in memory. If you used the `xclass` pattern and that class isn't in memory, Sencha Touch will require it on demand via the loader system, which could cause a performance blip in your app while the network request is going on for the required resource. The best way to avoid this is to define your plug-ins as required classes in your application. In case you're wondering, we'll cover dependency injection in relationship to applications in chapter 11.

The theory of plug-ins is rather simple, but in order to fully understand how this stuff works you'll have to put it into practice.

Figure 10.8 Your plug-in lets you collapse a fieldset by tapping its title.

10.4.2 Developing your plug-in

The plug-in allows users to expand and collapse fieldsets by tapping the fieldset title. Figure 10.8 shows what the plug-in looks like rendered on screen. The collapse of the fieldset is animated, adding a bit of flare to the plug-in.

You'll call this plug-in FieldsetCollapser, and you'll have to create js/Filedset-Collapser.js in your project. The class is much too large for print so we'll present it in three phases. We'll begin by looking at the base class structure and an applier method. Then, we'll look at the largest method, `init`, followed by the last three methods for the class. The next listing shows the base class structure.

Listing 10.15 Base structure for the FieldsetCollapser plug-in

```
Ext.define('Ext.ux.FieldsetCollapser', {          ❶ Defines FieldsetCollapser
    extend : 'Ext.Component',
    alias  : 'plugin.fieldsetcollapser',          ❷ Sets plug-in alias

    config : {                                    ❸ Defines config object
        collapsed        : false,
        body             : null,
        indicatorEl      : null,
        initialCollapse  : false,
        bodyHeight       : null
    },

    applyCollapsed : function(collapsed) {        ❹ Adds applier method
        var me = this;
```

```
        if (collapsed && ! me.getBody()) {
            me.setInitialCollapse(collapsed);
        }
        else if (me.getBody()) {
            me.doCollapse(collapsed);
        }

        return collapsed;
    },

    init : function(parent) {},

    onTitleTap : function() {},

    doCollapse : function(collapsed) {},

    updateIndicator : function(val) {}
});
```

The base definition for the `FieldsetCollapser` class ❶ in listing 10.15 is rather busy. To provide lazy instantiation you have to set the alias as `'plugin.fieldsetcollapser'` ❷. You're extending `Ext.Component` because it provides the necessary `config` object integration that we've come to love.

The `config` object ❸ contains a lot of custom properties. To learn how they're used see table 10.2.

Table 10.2 The plug-in `config` object properties

Property	Purpose
`collapsed`	This property is used as a configuration option. Implementations of this plug-in can be set with the value as Boolean `true` or `false`. A `true` value will ensure that the fieldset is rendered collapsed. This method has a special applier method that we'll cover.
`body`	This is a private member used to cache the reference to the fieldset's `body` element.
`indicatorEl`	Use this to store a reference to the element that indicates whether the fieldset is collapsed or expanded. You'll render a plus sign (+) to show that it can be expanded and a minus sign (-) to show that it can be collapsed.
`initialCollapse`	This property is used internally to specify whether you want to collapse the fieldset as soon as it's rendered. Because of the timing of how things get painted in the DOM, you have to work a bit to keep track of certain steps in the rendering process.
`bodyHeight`	Because you'll be collapsing and expanding the fieldset `body` element you'll need to keep a copy of the original height so that when the fieldset is collapsed you can expand it.

`applyCollapsed` ❹ is an applier method for the collapsed `config` object. You use the applier pattern for a few reasons:

1. You want to allow for the use of the automatically generated `setCollapsed` instance method. As you'll see in the `init` method you'll bind a `setCollapsed` method plug-in's parent (`Fieldset`).
2. You want to allow the fieldset to render collapsed initially. This means that when the plug-in is implemented, `"collapsed : true"` is passed. If you didn't have an applier function you wouldn't have an opportunity to collapse the fieldset after it's rendered.

The applier function takes care of these issues by checking to see if `collapsed` is `true` and whether the fieldset `body` element exists. If `collapsed` is `true` and the fieldset `body` doesn't exist, then you must cache the `collapsed` value via the automatically generated `setInitialCollapse` method. If `collapsed` is `false` and the `body` element is present, then you call upon `doCollapse`, invoking the animation of the fieldset's `body` element height change.

Next we'll look at the rather large `init` method for your plug-in, as shown in the next listing.

Listing 10.16 The `init` method for the FieldsetCollapser plug-in

```
init : function(parent) {
    var me          = this,
        title       = parent.down('title'),
        body        = parent.element.down('.x-dock-body'),
        titleEl     = title.element,
        bodyStyle = body.dom.style,
        indicatorEl,
        bodyHeight;

    setTimeout(function() {                                      ❶ Defers height
        bodyHeight = body.getHeight();                             calculation
        me.setBodyHeight(bodyHeight);

        if (me.getInitialCollapse()) {                           ❷ Collapses body
            bodyStyle.height = 0;                                  element
            me.updateIndicator('+');
        }
        else {
            bodyStyle.height =  bodyHeight + 'px';
        }

    }, 1);

    titleEl.on({                                                 ❸ Registers tap
        scope : me,                                                event
        tap   : 'onTitleTap'
    });

    indicatorEl = Ext.Element.create({                           ❹ Creates
        style : "font-weight: bold; float: left; width: 16px;"     new element
    });

    indicatorEl.setHtml('-');                                    ❺ Sets initial
                                                                   state
```

```
        titleEl.down('.x-innerhtml').insertFirst(indicatorEl);     <-- 6 Injects indicator

        me.setIndicatorEl(indicatorEl);
        me.setBody(body);

        body.setStyle({
            '-webkit-transition-property' : 'height',
            '-webkit-transition-duration' : '.5s'
        });

        parent.setCollapsed = Ext.Function.bind(me.setCollapsed, me);
    },
```

In listing 10.16 the `init` method does quite a bit for this plug-in and has some of the most advanced use of Sencha Touch that you've seen in this book. After creating a number of local references you have a function that's executed after 1 ms via a `setTimeout` ❶ method. Inside `setTimeout`, you get the current calculated height of the fieldset `body` element. You cache it via the autogenerated `setBodyHeight` method and then test to see if your plug-in was configured initially with `collapsed` set to `true` ❷ (remember `applyCollapsed`). If that value is set to `true` you immediately set the height of the fieldset `body` element to 0, forcing it to collapse immediately. You update the indicator to show that the fieldset can be expanded. If the plug-in isn't initialized with `collapse` set to `true`, then set the fieldset `body` element's `style.height` attribute to the calculated height. You do so because animations don't work well when the height isn't explicitly set (it's set to `auto` by default). So in order to prevent a glitch in the animation you must perform this step.

Why use setTimeout?

The reason you use the `setTimeout` function is that at the time the `init` method is called the fieldset's `body` element hasn't been injected into the DOM, preventing you from getting the proper height or even setting it. The deferred execution allows the DOM to be painted, and this logic to be executed soon thereafter, ensuring that you can do everything you need to do.

After the `setTimeout` function you register a `tap` event listener (`onTitleTap`) for the fieldset `title` element ❸. You'll learn about that method in the next listing.

Next you create an element for the collapse status ❹, set its initial HTML to `'-'` ❺, and then inject it into the fieldset `title` element's DOM ❻ structure. After the injection you cache references to the indicator and the fieldset `body` element.

The final bit of work that this method performs is to set some necessary CSS3 transition styling on the fieldset `body` element. These properties will instruct the element to animate the height of the element when it's changed and complete the animation within half a second. You also bind a `setCollapsed` API method to the `Fieldset` instance, allowing implementation code to collapse/expand a fieldset via a simple method call on the fieldset itself without having to dig for a reference to the plug-in instance.

That does it for the init method. The last three methods, shown in the next listing, are small potatoes compared to the monstrous init method.

Listing 10.17 The final three methods for the FieldsetCollapser plug-in

```
onTitleTap : function() {
    this.setCollapsed(! this.getCollapsed());          ❶ Toggles collapse/expand
},

doCollapse : function(collapsed) {
    var me          = this,
        body        = me.getBody(),
        bodyStyle   = body.dom.style,
        origHeight  = me.getBodyHeight(),
        indicator   = '-',
        height      = 0;

    if (collapsed) {
        indicator = '+';
    }
    else {
        height = origHeight + 'px';
    }                                                  ❷ Collapses, expands body element

    bodyStyle.height = height;
    me.updateIndicator(indicator);
},

updateIndicator : function(val) {                      ❸ Updates indicator element
    this.getIndicatorEl().setHtml(val);
}
```

As listing 10.17 shows, the last three methods are responsible for some core features but their methods are rather simple. onTitleTap ❶ is called when the fieldset title element is tapped, and it'll call this.setCollapsed(), passing in the negated result of this.getCollapsed(). The autogenerated setCollapsed method will call the applyCollapsed applier method. The applier method will call doCollapse to collapse or expand the fieldset body element.

doCollapse will collapse or expand the body tag ❷ based on whether the passed collapsed argument is true or false. It also updates the indicator element text accordingly via a call to updateIndicator ❸.

Look at that! You're done with sculpting the plug-in. It's time to work on the implementation to wrap up this chapter. The next listing may look lengthy but it's rather simple. Most of this stuff you've seen before.

Listing 10.18 Implementing your FieldsetCollapser plug-in

```
Ext.Loader.setConfig({
    paths : {
        'Ext.ux' : 'js/'
    }
});
```

```
Ext.require([
    'Ext.MessageBox',
    'Ext.form.Panel',
    'Ext.ux.FieldsetCollapser'                    <-- ❶ Requires plug-in
]);

Ext.Viewport.add({
    xtype  : 'toolbar',
    docked : 'top',
    title  : 'Collapsible Fieldsets'
});

Ext.Viewport.add({
    xtype  : 'formpanel',
    items  : [
        {
            xtype       : 'fieldset',
            defaultType : 'textfield',
            title       : 'Name info',
            plugins     : {                        <-- ❷ Expands plug-in by default
                type : 'fieldsetcollapser'
            },
            items       : [
                {
                    label : 'First'
                },
                // more fields here
            ]
        },
        {
            xtype       : 'fieldset',
            defaultType : 'textfield',
            title       : 'Address info',
            plugins     : {                        <-- ❸ Configures plug-in in collapsed state
                type      : 'fieldsetcollapser',
                collapsed : true
            },
            items       : [
                {
                    label : 'Address'
                },
                // more fields here
            ]
        }
    ]
});
```

The implementation for the plug-in is rather simple. You first ensure that the plug-in is required into the namespace ❶. Then you create the first fieldset for the Form panel configuration. There you set plug-ins to a simple lazy-instantiation object whose `type` property is set to the same `'fieldsetcollapser'` ❷, your plug-in alias. There's a second fieldset that you configure. It also has an instance of our FieldsetCollapser plug-in ❸, but it's configured to be collapsed initially by default.

Figure 10.9 Your plug-in in action

Figure 10.9 shows that your plug-in works as you expect, with the second fieldset collapsed initially.

You just saw that with a little bit of work, you can create a plug-in to inject functionality into a standard Sencha Touch component. The truth is that in order to create this plug-in we had to inspect the DOM and figure out how the fieldset was rendered. After this discovery we had to generate the plug-in to modify the fieldset's underlying HTML to allow it to be expandable or collapsible.

This concludes the exploration of plug-ins. You now have the basic knowledge to create plug-ins that can enhance functionality in your projects. If you have an idea for a plug-in and aren't sure if it's been done before, visit us at the Sencha Touch forums (http://sencha.com/forum). An entire section is dedicated to user extensions and plug-ins, where fellow community members have posted their work, some of which is completely free to use.

Sencha has a nice collection of user extensions that you can download at its own marketplace. You can visit it at http://market.sencha.com.

You now understand the mechanics of developing an extension to the framework and constructing a plug-in to meet real-world requirements.

10.5 Summary

In this chapter you learned how to implement the prototypal inheritance model using the basic JavaScript tools. You saw how this inheritance model is constructed step by step. Using that foundational knowledge you refactored classes using the `Ext.define` class definition method.

Next you took that foundational knowledge and applied it to the extension of a `List`. You were able to inject custom functionality to slide in custom DOM that was injected by your extensions. You also took the opportunity to fire custom events based on the configuration that was passed to your `ActionList` extension.

Finally you created a plug-in that allowed you to collapse and expand fieldsets via a tap of the `title` element. While doing this you learned how to inject custom DOM and CSS attributes to make the animation of the expansion or collapse of the fieldset a snap.

In the next chapter we're going to look at how you work with the SDK tools to develop an application for deployment. Afterward, we'll look at creating a simple application that will work for phones and tablets via Sencha Touch profiles.

11 Building Sencha Touch applications

This chapter covers

- Exploring the app development cycle
- Using Sencha Cmd
- Deploying an example app
- Using Sencha Microloader

Until now, you've focused on pieces of Sencha Touch, learning how they work and how to use them properly. In this chapter we're going to show you design patterns that'll help you develop your applications from start to finish. It's impossible for us to cover every facet of application development, so we'll center on many of the key things that you'll need to know.

This chapter is structured differently than what you're used to. The workflow is tutorial-like, and you'll progressively build on things you've learned. We realize that you might be in a hurry to develop your applications and that you might need a simple answer to a question such as "How do I bind controllers to custom view events?" But be aware that not reading this chapter as it was written may lead to confusion, because we'll reference prior material as we progress through the chapter.

You'll start by looking at a process that governs how applications are typically developed. From there you'll explore Sencha Cmd and use it to generate your application template project. We'll walk you through many basic concepts and show you how to use device profiles to create a single application that can service phones and tablets. You'll also learn how to fire custom events from your views and listen to them via Touch controllers.

After you've developed your application you'll use Sencha Cmd to create testing and production builds. There will be a lot of discussion of the Sencha Touch Microloader, and you'll learn how it plays a key role in every phase of application development and deployment.

This is the most intense, yet rewarding, chapter of this book. Let's go and have some fun!

11.1 The Sencha 30,000-foot view

Since MVC was introduced applications developed with Sencha Touch have to follow specific design patterns both in the JavaScript space and inside the filesystem. There's also a general process that you must follow to ensure a smooth development cycle with minimal tripping over your own feet.

If you're a seasoned developer you probably know all of this stuff, but if you're relatively new to application development and deployment design patterns understanding this workflow is crucial to the success of any project.

11.2 Typical application development workflow

The application development process is detailed in the (relatively) simple workflow shown in figure 11.1. We'll walk you through it quickly.

We made the development workflow diagram simple enough for just about any developer to digest. It illustrates a generic workflow loop that you'll go through when developing applications with any product. Table 11.1 breaks it all down.

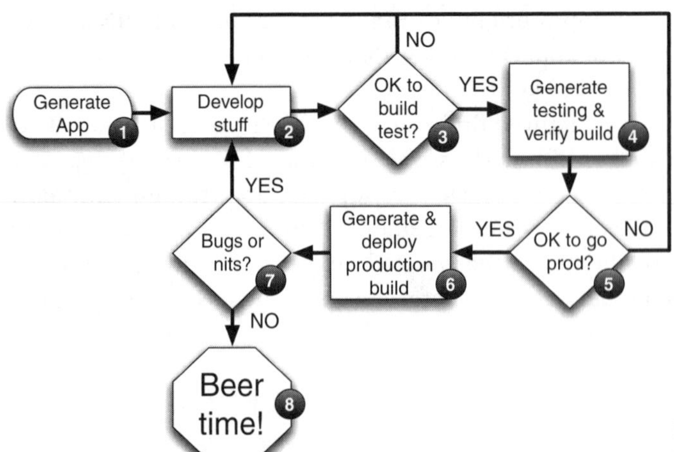

Figure 11.1 A generic application workflow diagram

Table 11.1 The application development cycle

Step	Process explanation
1	You begin with the generation of the application. As you'll learn later in this chapter, you do so using the Sencha Cmd toolset; all of the basic directory structures will be generated for you and you're only responsible for developing your application.
2	This step is where you spend most of the application development time generating model, store, view, and controller classes. In this process you can develop custom themes and fix bugs. Any and all changes are subject to being tested, which takes place in step 3.
3	At some point you must decide to show off what you've developed to your customers or QA team members. This step marks this decision. If you're not confident that you're ready to move forward, jump back to step 2.
4	Once you're confident that your application can be viewed by QA staff or customers you use Sencha Cmd to generate a testing build and deploy it to your designated environment. A testing build is a concatenation of your JavaScript files and you bundle it with other assets like CSS and images. A testing build is generally deployed to a testing environment for QA and/or customer approval.
5	After you've deployed your testing branch determine if it's all right to go to production. Many decisions can lead to the NO branch of this process, which can include bugs and missing features that have yet to be developed. If for whatever reason you've reached the NO decision you must return to step 2 and work your way forward until you're confident that you can move ahead with a production build.
6	The production build process is very much like doing a testing build, except your JavaScript is run through a minification process (in which the code size is reduced) and the Sencha Touch Microloader is enabled for code caching. (We'll cover the Microloader in much greater detail later in this chapter.) This is the code that'll be shipped off to your production environment for customer use. The greatest stress of an application development process is generally experienced between this step and the next.
7	Sometimes you go to production with bugs. Software tends to be among the most complicated things that humans generate (even in JavaScript!), and your software will be shipped with bugs. It's inevitable! Shortly before or after a production launch you need to identify bugs in your software. If you find bugs return to step 2 and work through the process again until you do another production push. At this point in the game you might want to jump from step 2 to step 6. But skipping steps can lead to what are known as regression bugs (breaking things that worked in the past), so it's wise to work through each step in the process even though it may be time-consuming and painful. The last thing you want is your boss complaining that you broke something that used to work because you cut corners.
8	Congratulations! At this point of the game no known bugs are found, so it's time to relax and have a drink.

The development workflow in table 11.1 works very well for most cases and we suggest you use it for future projects. You need discipline to be able to follow it, which is something that the greatest developers have, no matter what the programming language or technology.

As we move forward through this chapter we'll walk you through the major development steps, beginning with a quick look at the toolset to kick things off: Sencha Cmd.

11.2.1 What is Sencha Cmd?

Don Griffin's engineering team at Sencha developed the successor to SDK Tools 2.0 and he renamed it Sencha Cmd 3.1. This all-new and more robust version of the Sencha command-line utilities has been rebuilt from scratch using Java as the core technology to aid in cross-platform reliability as well as speed. It comes with optional integration with Apache Ant for those of you who use it for your continuous integration build processes.

So why use Sencha Cmd? We could probably write a few more pages on why you'd want to use Sencha Cmd, but table 11.2 presents a quick list of features that we hope will whet your appetite.

Table 11.2 Sencha Cmd features

Feature	Explanation
Multiple platforms supported	Sencha Cmd will run on your Windows, Mac, and popular Linux (32- or 64-bit) systems with ease. This means that you can develop your code in one platform and rely on another to execute your production or test build processes accordingly.
Code-generation tools	Sencha Cmd can be used to start your application and concatenate/minify your application. It can also be used to generate code stubs for controllers, views, models, forms, and profiles. Using Sencha Cmd can assist with code consistency when more than one person is working on your application.
Awesome documentation	Part of what makes this toolset so great is its amazing documentation. It begins with an introduction of what Sencha Cmd is all the way down to a quick reference for the toolset.
Great command-line help	If you get stuck on a particular feature of Sencha Cmd and need help, the toolset comes to the rescue by offering help in your shell via the `help` flag.
Multiple build targets	With the Sencha Cmd toolset you can develop testing, production, and native builds. With testing, your code is concatenated but not minified, so it's readable and thus possible to debug. The production build is basically a testing build with the addition of code minification and a few more features to allow your apps to bootstrap lightning-fast after the first load.
Integration with UglifyJS, Google Closure Compiler, or YUI Compressor	JavaScript is an interpreted programming language. Often you write programs to be human-readable. This means lots of whitespace, comments, and so forth. All of that stuff has to be parsed out by the browser's JavaScript engine. Doing so is time-consuming, but one way to speed things up is to remove white space and change variable names and function arguments to single letters. Another way is to look at the JavaScript and reorganize things to make it easier for your code to be digested by the browser's JavaScript engine. Each tool has its benefits and drawbacks, all of which are beyond the scope of this book. Using any one of these tools is way better than using none at all!

Table 11.2 Sencha Cmd features *(continued)*

Feature	Explanation
Native packaging	Sencha Cmd can take your Touch apps and wrap them in a native packager, allowing you to easily distribute in the Apple App Store or Google Play.
Automation hooks with or without Ant	For those of you who rely on Apache Ant to deploy your applications, you can integrate Sencha Cmd with your Ant tasks. If you don't use Ant you can still generate builds using the command-line utilities; they'll throw exceptions and exit with an unfavorable return code (anything greater than 0) so your automation scripts and tools can catch the exception and deal with it accordingly.

You've just completed our 30,000-foot view of Sencha Cmd, and it's time to start some preparation work. Before you start developing your application let's discuss where to get Cmd and some of its dependencies.

11.2.2 Obtaining Sencha Cmd

Sencha Cmd uses a suite of technologies that it installs on its own but it requires you to have dependencies already on your system. Here's a quick list of what's required. Chances are you might have some of these dependencies already on your development environment and installation can be done in minutes on a fast internet connection:

- Java JRE 1.6: www.oracle.com/technetwork/java/javase/downloads
- Ruby on Rails: http://rubyonrails.org/
- Sass: http://sass-lang.com/download
- Compass: http://compass-style.org/
- Sencha Touch 2.2 or greater: www.sencha.com/products/touch

Obtaining Sencha Cmd is as easy as visiting http://sencha.com/products/sencha-cmd and then downloading and installing it. We're not going to go over the process because Sencha has a nice install guide on their website for each system.

You should already have a copy of Sencha Touch, but if it's version 2.0.2 or lower you'll need to upgrade. The easiest way to check what version of Sencha Touch you have is to look at the release-notes.html file in your browser. You'll find the version at the top of the rendered page.

The stage is now set for you to start developing your application.

11.3 Creating your application container

To kick things off let's invoke step 1 of the development process (figure 11.1). To generate your Sencha Touch application you'll need to drop down to a command shell. Listing 11.1 contains the command-line output of the application generation process.

> **A quick word about OS and syntax differences**
> We'll demonstrate using Mac OS X. The commands in Linux will be the same but the Windows counterpart will require a simple change to how the paths are written. Unix-style paths look like this: /path/to/your/project. Windows paths look something like this: C:\path\to\your\project. We'll leave it up to you to make the proper substitutions and changes where required.

Listing 11.1 Generating an application

```
jay:~ jgarcia$ cd /www/touch/sencha-touch-2.2.0/          <- 1 Makes changes to Touch SDK directory
jay: sencha-touch-2.2.0 jgarcia$ sencha generate app App /www/myapp   <- 2 Instructs Cmd to create app stub
Sencha Cmd v3.1.1.274
[INF]
[INF] init-plugin:
[INF]
// Truncated
[INF]      [mkdir] Created dir: /www/myapp/packages
[INF]
[INF] copy-framework-to-workspace-impl:
[INF]       [copy] Copying 1976 files to /www/myapp/touch        <- 3 Copies Touch resources
[INF]       [copy] Copied 233 empty directories to 1 empty directory under
                    /www/myapp/touch
[INF]       [copy] Copying 1 file to /www/myapp/touch
[INF]       [copy] Copying 1 file to /www/myapp/touch
[INF] [propertyfile] Updating property file:
                    /www/myapp/.sencha/workspace/sencha.cfg
// Truncated
[INF] generate-starter-app:                                       <- 4 Creates app stub
[INF]      [mkdir] Created dir: /www/myapp/app/model
[INF]      [mkdir] Created dir: /www/myapp/app/controller
[INF]      [mkdir] Created dir: /www/myapp/app/store
[INF]      [mkdir] Created dir: /www/myapp/app/profile
// Truncated
```

> **You don't have to follow our lead on file location**
> We're placing our files in /www/myapp for simplicity but you can put your project wherever you feel most comfortable. Also be sure to have your local web server up and running. Your web server will also need access to the files you've just created with Sencha Cmd.

Your first step is to change the directory to wherever you put your extracted package contents for Sencha Touch 2.2 (or greater) ❶. We like to place everything we do in /www/touch for ease of typing but you don't have to follow our pattern.

NOTE The output from Sencha Cmd we're showing you is generated by an early copy of the tool. The results you see may differ slightly.

Next you instruct Sencha Cmd to generate your application stub ❷ via the following `sencha` command construct:

```
sencha generate app App /www/myapp
```

The first argument to the `sencha` command invokes the generate mode. There are two arguments to the app generation command. The first is your application JavaScript namespace (`App`) and the path to where the files are going to be placed.

> **NOTE** It's worth noting that everything you do with Sencha Cmd literally begins with `sencha` in the shell.

If you examine the output of the `generate app` command you'll find that it copies a lot of resources to your destination, /www/myapp. First is a copy of the Sencha Touch library ❸ minus the fluff (docs, examples, etc.) as well as the Sass source files and images. You also see output where an app/ directory is created ❹ with subdirectories for supporting classes.

To better understand what's happened let's look at the directory and talk about it a little.

11.3.1 Examining Cmd app resources

You just generated your stub app and saw a flurry of output from Sencha Cmd. Before you can start building your application you need to figure out what was laid out. Take a look at figure 11.2.

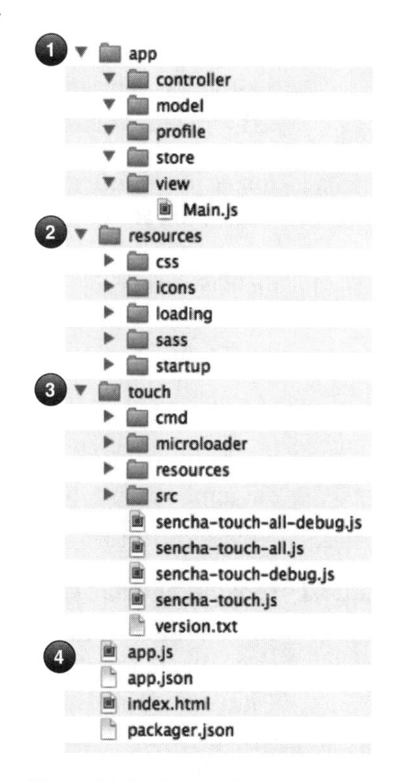

Figure 11.2 A snapshot of your newly generated application

The first thing to look at is the app directory ❶. You'll find directory stubs for your controller, model, profile, stores, and views. In the view directory you'll find one `Main.js` class. Feel free to open that file in your editor. You'll see that this file contains a simple tab panel implementation with a "Welcome to Sencha Touch 2" set of screens. You'll be overwriting this file.

Next look at the resources directory ❷. This directory contains the necessary directories that contained your compiled CSS, launch icons, custom theme stub (Sass), and loading screens (startup). The loading directory is a legacy directory and will most likely be removed from the distribution soon.

The touch directory ❸ is a directory that contains all of the Sencha Touch resources. Generally you shouldn't be entering this directory at all.

The last group of files ❹ plays a role in your application's development life cycle. We're going to talk about these files from a high level and do deep-dives later on:

- app.js is the file that defines the structure of your application. It's designed to include all immediate dependencies—generally controllers and profiles. You can add any other class to be required in your app.js. Take a look at it in your editor and get some exposure. We'll look at it in greater detail later on.
- app.json is a file that describes to Sencha Touch what resources to load in development mode and describes to Sencha Cmd how your application should be built. If you have external JavaScript code or libraries that you need to load with your application you can specify them in the "js" section of this file. We generally place them in app/lib/<library_name>/.
- index.html is used to load your application in the browser. It contains necessary things like a CSS3 boot animation and what's known as the Sencha Touch Microloader, which contains the minimum amount of JavaScript to kick everything off.
- packager.json is a file that's used by Sencha Cmd to package your application; it wraps your app in a native binary for deployment.

Be careful when modifying app.json!

It's absolutely normal and a sanctioned act to modify app.json to add your own resources or change the default configurations for how your application is bundled. Just be careful not to inject errors into the JSON file. Errors in this file will prevent your application from loading at all.

You've just finished generating your application stub. But before you start developing take a look at figure 11.3 to see what renders onscreen.

As you can see, the application stub generated by Sencha Cmd renders perfectly in the iOS simulator (and Chrome or Safari). This means that all your preparation work is done for your application development. You can begin breaking ground and start developing your app.

11.4 A view of what you're building today

You just generated the app stub via the Sencha Cmd toolset and saw what the stubbed-out application looks like in a browser. You're going to start developing your application soon, and we'll begin by discussing exactly what you're building.

11.4.1 Looking at what you're building

To exercise a lot of what you've learned and to implement new concepts you're going to develop a relatively simple application that works on tablets and phones via Sencha Touch profiles. The heart of the application is simple. It's a contact editor with two screens: a list of contacts and a detail screen that lets you modify the data. Most of the work that you'll do involves writing the controller logic that manages the interaction models between phone and tablet. Let's look at the phone version first (figure 11.4).

The point of the example application is to see data and modify it. That's it! Figure 11.4 demonstrates the two main screens that you'll implement on a phone. The navigation is simple. Tap on an item in the list (left) and slide in the detail screen (right). The back button returns the user to the list and commits the changes to the list.

The tablet version is a bit different. Let's take a quick look at it (figure 11.5).

As illustrated in Figure 11.5, in the tablet version of the application you'll be displaying both views. When the user taps on an item in the list the form in the center updates with the relevant data. You'll notice that there's a save button in the tablet version of the detail screen, whereas there's no save button in the phone version (figure 11.4). To persist the data on the tablet version the user will have to hit the save button.

You now have an understanding of what your app will look like when you render it on both phone and tablet devices. Next let's take a quick glance at the namespace you'll be developing.

Figure 11.3 The application stub generated by Sencha Cmd rendered on the iPhone simulator

Figure 11.4 The example application on a phone

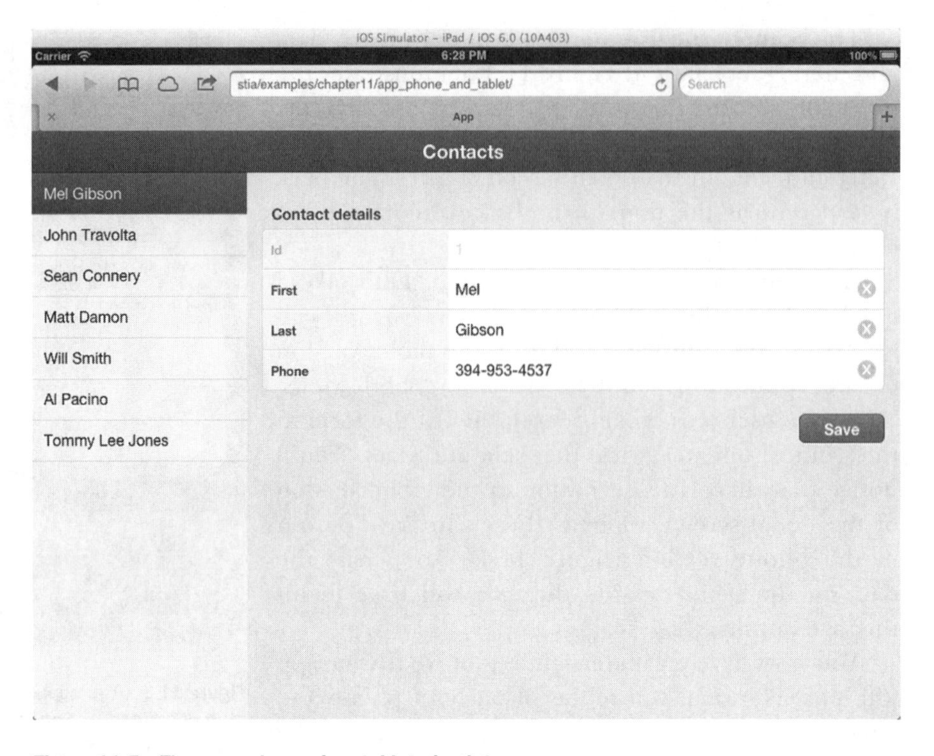

Figure 11.5 The example app in a tablet simulator

11.4.2 *A quick glance at the namespace*

Figure 11.6 illustrates how the namespace for your application will look once you're done programming.

The namespace for this application may seem insanely complicated at first glance, but in reality it's simple. Here's why.

> **File placement is important**
>
> It's extremely important that you place your classes on your filesystem exactly how they're organized in their namespace. As you develop this application we'll remind you what classes go where, but it'll be up to you to remember this rule when working on your own applications. Failure to do so might lead to exceptions thrown by Sencha Touch—and frustration.

Everything begins with the `App` namespace that you told Sencha Cmd to generate for you. There are subnamespaces (packages) defined under `App`, which are controller, model, profile, store, and view. Each of these packages contains either subpackages or classes.

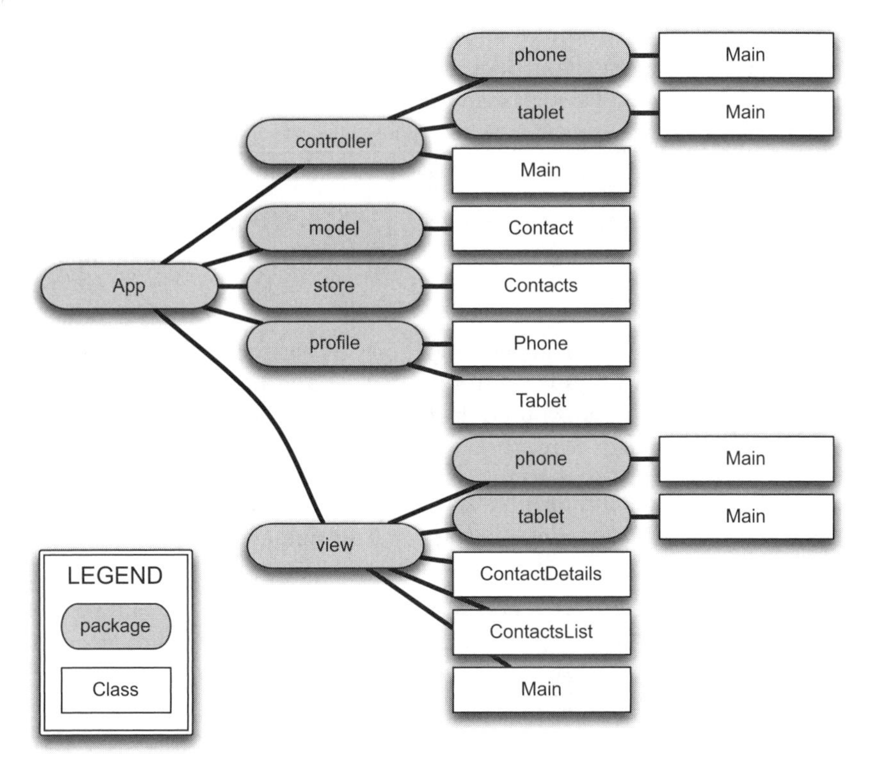

Figure 11.6 The application namespace that you'll create

> **Filename capitalization**
>
> We've been involved with projects that have had issues when migrating from one system type to another due to simple mistakes, where a class name and its related file were capitalized differently. Development environments or your filesystem and OS may tolerate these differences, but testing or production environments may not. So please try to be diligent when it comes to naming your files and classes. Your diligence will pay off! It's worth noting that using Sencha Cmd to generate your class files can assist with this effort.

If you look at the controller namespace you'll see that it has a `Main` class as well as two subpackages, phone and tablet, each with its own `Main` class. We'll dive further into this later, but to quell your curiosity, you define one main controller in the controller namespace that's used as a common class, and the Phone and Tablet profiles will subclass `controller.Main` to implement specific behaviors for the tablet and phone versions.

For the model and store packages you have a class each. Because the `Model` and `Store` classes won't need to change for the tablet or phone versions you have no need to make a subpackage.

NOTE Since a `Model` instance is a single instance of a `Contact` you name it in singular, whereas the `Store` instance deals with one or more `Contacts`, so you name it in plural. This is a common convention for naming `Model` and `Store` classes.

The profile package is used to define device profiles to Sencha Touch. In this case you're only dealing with two types of devices, and as you'll see when you implement the code, you're going to be general in how you define what a tablet or phone is to your application.

NOTE With Sencha Touch device profiles you can get extremely granular and create specific profiles for different devices and name them exactly how you want to. For example, you could have one profile for iPhone4, another for iPhone5, another for SamsungGS3 (Galaxy S3), and so forth.

Lastly, the view package contains packages for the profiles, a generic main view, and two views that begin with the word Contact. You named the class files (and classes themselves) in such a way that they describe what they do. For example, ContactsList is a list of contacts and ContactDetails shows details of a contact. When you name your classes keep in mind what exactly the class is trying to do and optionally append to it the superclass's name or something that describes what it does.

That wraps up all of the preliminary work you have to do to develop the example application. Next you'll grab your shovel and start digging in!

11.5 Building the Phone profile version of your application

You'll develop this application in phases. These phases are all encompassed by step 2 of our development process, and you'll stay in this step until the views are done and the interaction models are all wired up.

Here's how you'll develop this application beginning with the Phone profile:

1. Define the model and data stores.
2. Define the Phone profile.
3. Stub out the main and Phone profile main controller.
4. Define the main and Phone profile main views.
5. Wire up the interaction models for the Phone profile.
6. Define the Tablet profile controller.
7. Define the tablet view.
8. Wire up the interaction model for the tablet controller.

Once you're satisfied with how the app works on the phone and tablet you'll use Sencha Cmd to generate testing and production HTML5 builds.

11.5.1 Developing your data model and store

You're beginning with the data model and store because they'll be responsible for feeding data into the various contact views. Consequently, these two classes will be some of the simplest that you'll create. You won't see any of these classes in action until you start to develop the `ContactsList` class.

Because the model feeds the store let's start with that. Normally you'd write this class by hand but this is an excellent opportunity to show you how to use Sencha Cmd to generate a class file. Here's the Sencha Cmd syntax for creating the file app/model/Contact.js:

```
sencha generate model -n Contact -f firstname,lastname,phone
```

The `-n` flag allows you to define a class and file name, whereas `-f` lets you define all of the model's fields. Running the command yields the results shown in listing 11.2.

> **NOTE** If you want to learn more about the `sencha generate model` syntax simply ask Sencha for help by using the following command syntax: `sencha help generate model`. Remember that you can get help for any of the commands.

Listing 11.2 The `Contact` model class

```
Ext.define('App.model.Contact', {
    extend : 'Ext.data.Model',

    config : {
        fields : [
            {name : 'firstname', type : 'auto'},
            {name : 'lastname', type : 'auto'},
            {name : 'phone', type : 'auto'}
        ]
    }
});
```

❶ Defines Model class

❷ Sets up prototypal config object

The model definition is very simple. It begins by defining the model class namespace ❶ and extending `Ext.data.Model`.

> **NOTE** You're defining the model with the `App` namespace. Later on when we look at app.js we'll talk about how the `App` namespace is first bootstrapped.

Next you have the prototypal config object ❷, which contains the data fields that'll help drive the list and form views you'll create later on.

With the model defined you can next define your store. For simplicity you won't be using Sencha Cmd, but we'll show you the classes themselves. Create the file `app/store/Contacts.js` and put the content shown in the following listing inside the file.

Listing 11.3 The `Contacts` store class

```
Ext.define('App.store.Contacts', {              <- 1 Defines Contacts store
    extend   : 'Ext.data.Store',

    alias    : 'store.contacts',
    requires : ['App.model.Contact'],

    config : {                                   2 Uses contact model
        model : 'App.model.Contact',             <-
        data  : [
            {                                    <- 3 Throws in data
                id        : 1,
                firstname : 'Mel',
                lastname  : 'Gibson',
                phone     : '394-953-4537'
            },
            {
                id        : 2,
                firstname : 'John',
                lastname  : 'Travolta',
                phone     : '357-642-3162'
            }
            // More data in the downloadable example
        ]
    }
});
```

Listing 11.3 illustrates the `Contacts` data store extension (1). Within the class definition is an alias declaration which allows you to use shorthand notation to configure an instance of this store. The contact model is also set up as a dependency for this store so that when this class is defined the contact model is loaded and registered in the appropriate namespace.

You configure your store to use your contact model (2) and throw in some mock data for your view to consume (3). The supporting `Model` and `Store` classes were easy to complete.

> **NOTE** We removed a bunch of records from the listing that you've seen in the images of the sample application in order to reduce the size of the listing. Feel free to copy from the downloaded version of this exercise or create your own following the data design pattern.

Next we'll start work on the application views of the application structure.

11.5.2 Creating the generic main view class

Before you can start loading stuff in the browser you'll need to generate a main view. In Sencha Touch applications the main view is generally responsible for navigation transitions. But depending on the device being rendered and the application, the main view may just act as a static canvas to render bits of your view. Let's revisit the way the app is rendered on the phone versus on the tablet (figure 11.7).

Figure 11.7 Your application on a phone

In the case of the Phone profile the main view will be responsible for displaying the list, then transitioning in the ContactDetails view via the card layout and slide animation to display detailed data. At the same time that the animation is initiated you'll display a back button to let users return to the ContactsList. All of this stuff is happening within the main view. Essentially, the top-docked toolbar, ContactsList, and ContactDetails will be children of the main view.

Recall that the tablet version of this application is different (figure 11.8).

The tablet version of the main view won't need to transition items from left to right because you have a ton of real estate. But it'll remain a parent to the top-docked toolbar (minus the back button), ContactsList, and ContactDetails views.

In reviewing the application you see that there's a main set of features that you can implement. This is why you need to create a generic main view and then move on to create versions of the main view for the Phone and Tablet profiles. See the namespace diagram in figure 11.9 for clarification.

To create the common main view you'll need to open up the app/view/Main.js file and replace its contents with those in listing 11.4.

Listing 11.4 The common main view class

```
Ext.define('App.view.Main', {
    extend : 'Ext.Container',
    xtype  : 'main',

    config : {
        fullscreen : true,
```

❶ Defines common main view

❷ Makes view full-screen

```
        items : {
            docked : 'top',
            xtype  : 'toolbar',
            title  : 'Contacts'
        }
    },

    setTitle : function(title) {
        this.down('toolbar').setTitle(title);
    }
});
```

❸ Adds top-docked toolbar

❹ Sets toolbar title

The common main view class in listing 11.4 is simple but it contains a bit of reusable code that your phone and tablet versions will implement. First you register the

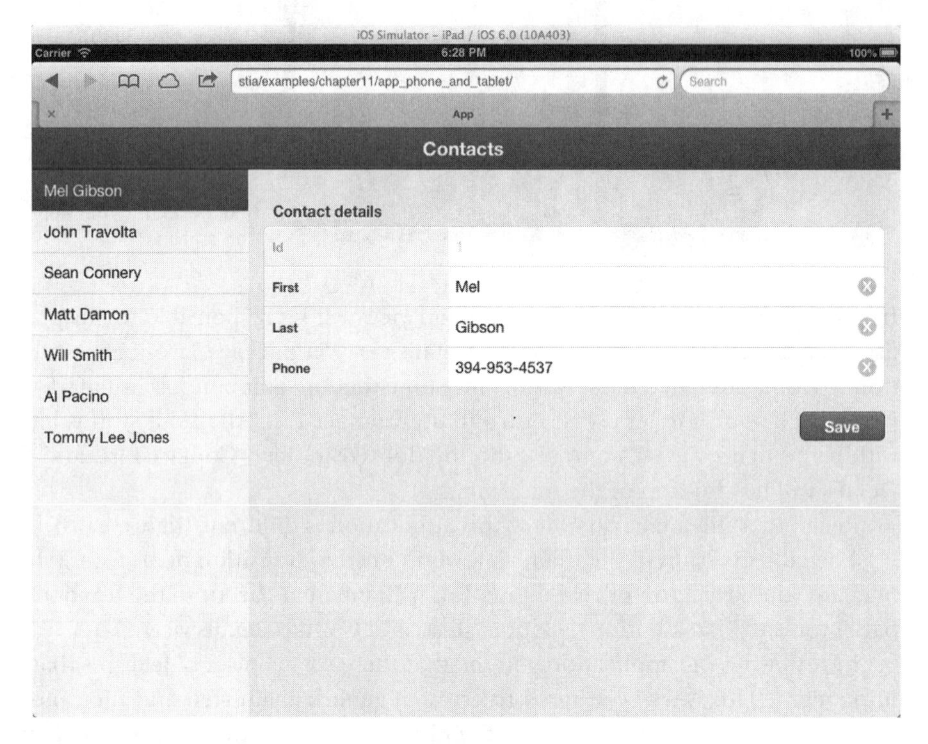

Figure 11.8 Your application rendered inside the iPad simulator

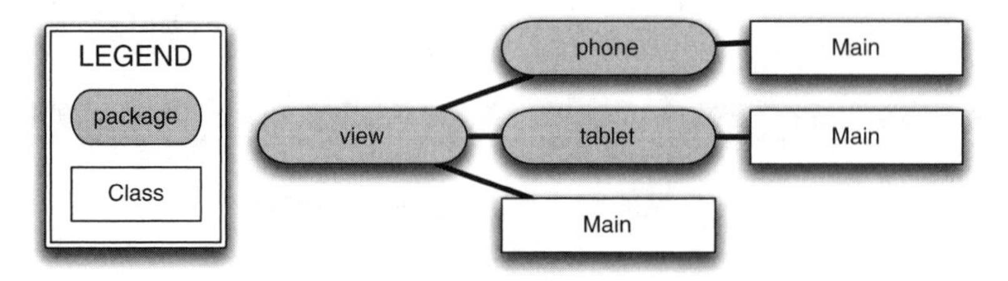

Figure 11.9 The Phone and Tablet profile-based main views will extend the common `view.Main` class.

App.view.Main class ❶, extending Ext.Container. You give it an XType of main which will be used in your controller to hook onto for events.

In the config section you instruct the view to be full-screen ❷ and configure a top-docked toolbar ❸.

Lastly, the setTitle utility method ❹ uses ComponentQuery to search for the toolbar and, by means of chaining, sets its title.

For a moment recall that you're doing the Phone profile first. You may be wondering, where's the layout? Where are the child items (such as ContactsList and ContactDetails)? As you'll soon find out, the Phone profile (and the Tablet profile for that matter) will be responsible for selecting which layout to implement and how to lay out the child items.

For now let's move forward with rendering the main view on the screen so you can soon start to view your incremental changes in the browser. To do so you'll have to update app.js to render your common main view. Afterward you can move forward with creating the ContactsList and Detail views and see them in action.

11.5.3 Looking at app.js for the first time

To generate your project you used Sencha Cmd, which created a bunch of directories and files for you. Earlier we discussed app.js, where we told you that it was used to describe and launch your application. Although you generally won't do a ton of work with app.js you must understand what's in it and what you're doing with it.

The next listing contains a version of app.js that we've modified a bit, but it still contains most of what's been generated by Sencha Cmd. We have to warn you, this is a long listing and we're going to enter a pretty lengthy and deep discussion here.

Listing 11.5 app.js

```
//<debug>
Ext.Loader.setPath({                    // ❶ Initializes class loader
    'Ext': 'touch/src',
    'App': 'app'
});
//</debug>

Ext.application({                       // ❷ Bootstraps application
    models: ["Contact"],

    name: 'App',

    requires: [
        'Ext.MessageBox'
    ],

    views: ['Main'],                    // ❸ Includes views

    icon: {                             // ❹ Defines app icons
       // Icons go here
    },

    isIconPrecomposed: true,
```

```
    startupImage: {                                    Starts image
       // startup images go here                    ❺ definitions
    },
    launch: function() {                                      Launches
        // Destroy the #appLoadingIndicator element        ❻ app
        Ext.fly('appLoadingIndicator').destroy();

        // Initialize the main view
        Ext.Viewport.add(Ext.create('App.view.Main'));    ❼ Executes
    },                                                       when app
                                                             is updated
    onUpdated: function() {
        Ext.Msg.confirm(
            "Application Update",
            "This application has just successfully been updated to the
 latest version. Reload now?",
            function(buttonId) {
                if (buttonId === 'yes') {
                    window.location.reload();
                }
            }
        );
    }
});
```

The app.js listing begins with the initialization of the Sencha Touch Loader system, where Sencha Touch itself is initialized via the Loader's `setConfig` method ❶. It also contains something that you probably haven't seen before:

```
//<debug>
 … some code …
//</debug>
```

These pseudo debug tags are used by Sencha Cmd to know what blocks of code to strip for production and testing builds. In this case, when you're in development mode you need the Loader system, but not in testing or production builds. This means that you can use these pseudo tags to wrap all types of debug messages or even include conditional breakpoints—as anything in between these tags will be stripped from your production builds.

The next important piece of this listing to focus on is the `Ext.application` method call ❷ where you pass an anonymous object—the app config object. This is the place that defines the bootstrapping of your application. A lot goes on here, so let's take a few moments to look further into it.

The property you find in the app config object is `models`, which is an array that contains the name of your recently created contact model. It's here because whenever you generate a class via Sencha Cmd it generates the associated keys for models, views, controllers, and profiles. At first glance this seems like an awesome feature, but don't use this pattern for a few reasons.

First is that the number of items—all of your views, apps, controllers, and profiles—can get rather large and unmanageable as your application grows. How we (and many other experts) use app.js is to bootstrap the application and potentially deal with application-level events. The requirements of views, models, and stores should always begin via profiles or controllers if you have no device profiles in your application.

Second, the way your required items are laid out may not be conducive to self-documenting code, which is something that the best programmers practice religiously. When you're done you won't be requiring models, stores, or application views from app.js. The rule of thumb we follow is that only global models, stores, utility classes, and overrides are required via app.js. Anything specific to a view should be required by that view or that view's controller.

So for now, remove the `models` key. We'll talk more about proper ways of thinking about dependency modeling a little later on.

Sencha Touch learns how to register your application namespace via the `name` key in the application config object. In this case your namespace is `App` (recall the Sencha Cmd command you executed to do this); therefore you can deduce that your application namespace can be accessed anywhere after your application is initialized. We'll look at this again when you render your app for the first time in your browser.

The next part is a `requires` array for `Ext.MessageBox`. This part was put in place by Sencha Cmd when you generated the application. The reason it's here is to satisfy the requirement of application update alerts (`onUpdated`), which we'll get to in a moment.

Next you'll find the `views` array ❸. Sencha Cmd also adds the `views` array when you generate your app for the first time. In this case it may make sense to leave it there, but to be honest, it's not going to be necessary, so you'll remove it. Once you get to the point where you're adding your profiles and controllers we'll show you the bigger picture as to how all of your dependencies get injected.

Moving forward you'll find `icon`, `iconPreComposed`, and `startupImage` keys. The launch `icon` ❹ and `startupImage` ❺ keys both are objects that describe image size and locations. We removed them from the listing to shorten it a bit. The `iconPreComposed` key is a Boolean used in iOS that basically tells it whether or not it should modify the icon. By default it's set to yes.

> **App icons and startup images**
> Sencha Touch comes with some well-composed launch icons (found in the extracted SDK in resources/icons) and startup screen images (resources/startup). These items are sized according to their need and are usable right out of the box, but they have Sencha's branding and will need to be replaced with your own branding before you move to production.

The last major sections of listing 11.5 are the `launch` method ❻ and `onUpdated` ❼. The `launch` method is called when the application instance is created but only after all dependencies are loaded. In your `launch` method the loading indicator is removed from the DOM. Later we'll discuss the loading indicator and why this step in the bootstrapping of an application is important. Next, an instance of `App.view.Main` is created and immediately added to the `Ext.Viewport` instance. This effectively renders the main view into your browser.

`onUpdated` is a function that's called by the Microloader in a production build only after an update to a production build of your application is found. In essence what this method does is display a message box telling users that the application has been updated successfully and give them an opportunity to reload the app or defer to see the changes after the application has been closed and reopened.

We're sure you have questions about the Microloader, production build, and MVC. We'll cover these in great detail later on. But for now we want you to see that what you've developed thus far does render on screen.

11.5.4 Checking in on progress

Figure 11.10 shows what your work has allowed you to do thus far. It's not much at the moment, but it's a great place to talk about the current progress.

In figure 11.10 we've expanded the `App` namespace in the debug tools to show you what Sencha Touch knows about your application. What you'll find is that there's an `.app` property of the `App` namespace. The `.app` property is quite literally a reference

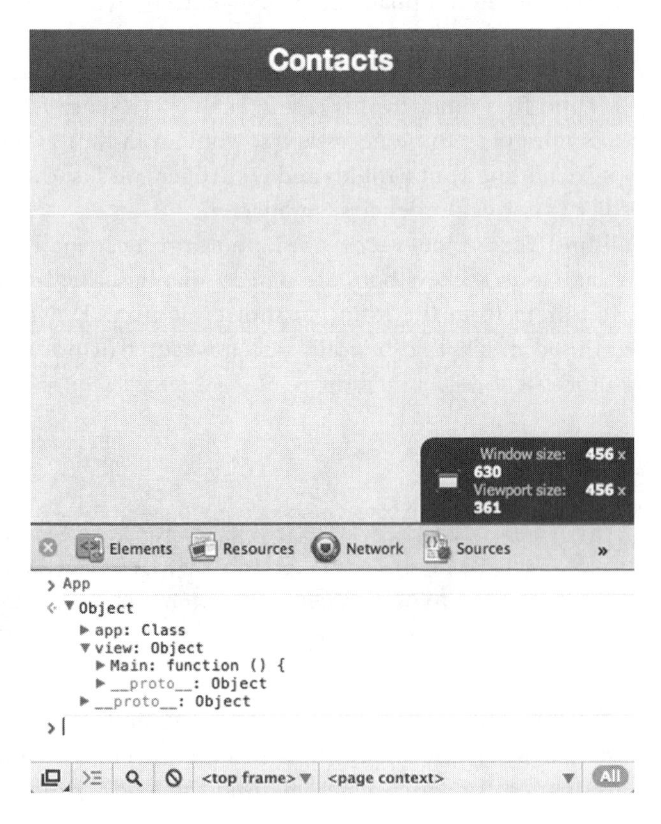

Figure 11.10 Viewing your app in a browser with its namespace visible in the debug tools

to the application instance. This means that within your code bits (controllers only, please!) you can access the application instance via `<ApplicationNamespace>.app` and can do things such as register event handlers or fire events.

Next look at the view subpackage, where you see a `Main` property. `Main` is the constructor for the `App.view.Main` common view that you see rendered in the browser's viewport.

The reason we're pointing out these things is so that you know that as you continue developing this application you can pop up the app in the browser, fire up the JavaScript debug console, and explore the namespace. Doing so can help you identify issues such as classes not being defined and namespace issues (remember, JavaScript is case sensitive). Exploring the namespace via the console can also allow you to get some visibility into some global utilities/objects that you may have defined.

You just bootstrapped an application with a basic custom main view. You have a lot more to do in this chapter, but now's a good time to take a detour that'll lead you to a understanding of how applications bootstrap with Sencha Touch. Once you have a good grasp of this concept you can build a controller and profiles, both of which are needed to complete your application using proper design patterns.

11.5.5 A quick lesson on how applications bootstrap

Application bootstrapping begins at index.html and typically ends with data being loaded. The following listing provides a truncated version of index.html. We removed the CSS to save space. If you want to read the CSS, feel free to open the file in your editor.

Listing 11.6 index.html (truncated)

```
<!DOCTYPE HTML>
<html manifest="" lang="en-US">
<head>
    <meta charset="UTF-8">
    <title>App</title>
    <style type="text/css">
       /* Loading indicator styles removed for sanity */
    </style>
    <!-- The line below must be kept intact for
                     Sencha Command to build your application -->
    <script id="microloader" type="text/javascript"
                   src="touch/microloader/development.js"></script>
</head>
<body>
    <div id="appLoadingIndicator">
        <div></div>
        <div></div>
        <div></div>
    </div>
</body>
</html>
```

index.html contains some stuff that's critical to getting your code loaded. It begins with the loading indicator CSS and its related HTML. The purpose of this HTML is to render something onscreen so the users can see that the app is doing something. In this case it's loading.

> **Remember `launch`?**
> Remember that inside the `launch` method (app.js) of your application definition is a step where the app loading indicator HTML is removed from the DOM. For convenience's sake, here it is: `Ext.fly('appLoadingIndicator').destroy();`.

In case you haven't seen the loading indicator yet figure 11.11 shows what it looks like.

The next thing that's important to look at is the `script` tag where the Sencha Touch Microloader JavaScript file is being loaded. As the comment above the line states, don't alter or remove this line. It's absolutely essential not only to the loading of your application but to the Sencha Cmd build phases as well. If you look at the tag you'll see that it's loading the development version of the Microloader.

If you browse to the touch/microloader/ directory you'll find that are three versions: development, testing, and production. Each version of the Microloader does something similar, except as you move from development to testing to production the loading process changes with each phase. As you'll learn later the production version of the Microloader kicks major butt! For now we'll focus on the application bootstrapping of the development version of the Microloader. As we move to testing and production we'll slow down to discuss how those work as well.

Figure 11.11 The application loading indicator rendered in a full-screen version of your application

> **Feel free to read the Microloader source**
> The Microloader source isn't minified so the source is readable. It's got some advanced stuff in there, but it's worth glancing at just to get some exposure to the code.

There's a lot of stuff that goes on to get your application bootstrapped. We're not going to delve into the gory details, but a general discussion will help you develop and debug Sencha Touch applications after you've put down this book. To help with this discussion we've created a simple workflow diagram, shown in figure 11.12.

Table 11.3 shows a quick rundown of how Sencha Touch applications bootstrap in development mode.

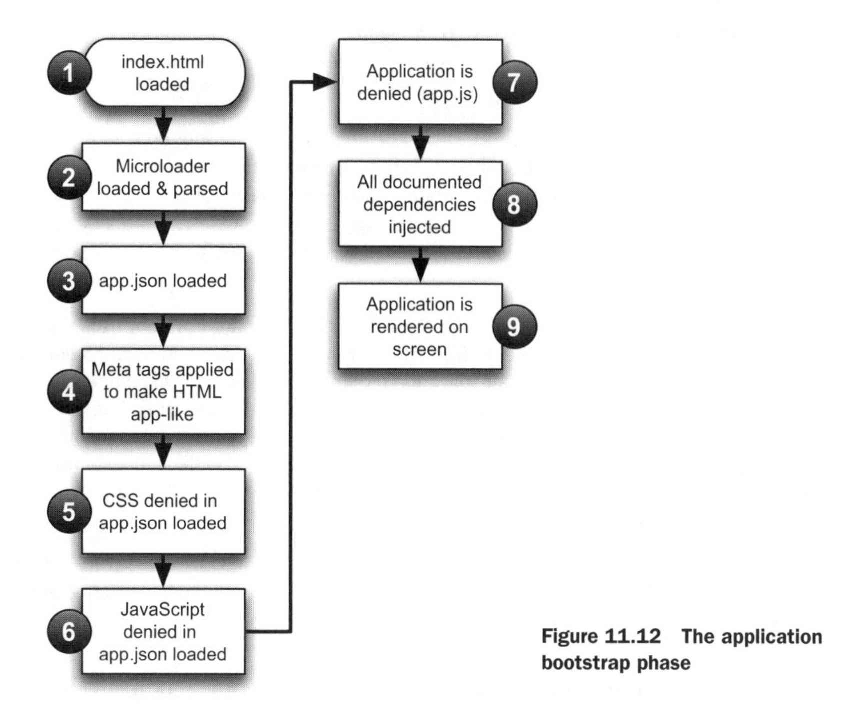

Figure 11.12 The application bootstrap phase

Table 11.3 The app loading process breakdown

Step	Description
1	index.html is first loaded by the browser. This is where the application-loading indicator is displayed.
2	The development version of the Microloader is loaded and parsed as part of index.html, but we broke it out as a separate step to highlight its importance. Steps 3–6 all occur within the Microloader itself.
3	From here the Microloader fires off a synchronous Ajax request to load app.json. Recall that app.json contains data objects that describe CSS and JavaScript resources to be loaded. In this case a single CSS file (resources/css/app.css) and two JavaScript files (touch/sencha-touch.js and app.js) are defined by default.
4	After the Ajax request completes, metatags are injected into the document to make the HTML document behave in an application-like way. These behaviors include the prevention of user scaling via the pinch gesture and the ability for the application to launch full screen (iOS only).
5	Next, all CSS files that were defined inside app.json are injected into the documented via link tags. The default is to have a single app.css (default Sencha Touch 2 theme) loaded, but you can add other CSS files to be loaded by the framework inside app.json. Just be sure to follow the design patterns and remember to not make mistakes in this file.
6	The next step is very important: this is where all of the JavaScript files that are defined in app.json are injected into the document via `script` tags. This includes the bootstrap file for the framework (sencha-touch.js) and your recently modified app.js file. This step kicks off a flurry of framework file loads to inject more of the Sencha Touch framework, including the event system, DOM utilities, layouts, component system, and a few other things. As we've discussed, sencha-touch.js contains the basics of framework (base utilities, loader, etc.) and the definition for the `Ext.application` method.

Table 11.3 The app loading process breakdown *(continued)*

Step	Description
7	Recall that inside app.js you call on `Ext.application` to define your app. A lot occurs inside the `Ext.application` method, but we'll highlight some parts you need to know. The first is that `Ext.Loader` is configured with your application namespace (`App`) pointing to what's known as the `appFolder('app')`. Next, `Ext.app.Application` is required, causing the base application classes to be loaded, including `Controller`, `History`, `Profile`, and `Router`.
8	In this next step documented dependencies are loaded by the Loader system. This step begins with `Ext.MessageBox` (remember that `requires` line in app.js?) and all of its dependencies. When you're done building this application profiles, controllers, views, models, and stores will be loaded in the order that they're documented via the `requires` array. This application dependency injection begins with profiles, followed by controllers. Views, models, and stores are loaded depending on how they're documented in the corresponding profiles and controllers. Keep in mind that profiles are optional and only need to be used when you want to develop an application beyond one device.
9	The last step to this phase is the application being rendered onscreen. This is when the application instance's `launch` method is called, removing the application-loading indicator and rendering the common main view class that you created earlier.

As you just saw, a lot's going on under the hood to launch your Sencha Touch applications. This discussion has paved the way for you to continue constructing your application without the fog of the unknown when it comes to application bootstrapping.

We'll continue with the addition of a Phone profile and related controller.

11.5.6 Adding the Phone profile

So far all you've done is generate an application stub and modify it so you have a custom main view displaying. Next you'll need to add a Phone profile, which will call on the phone-specific controller to manage the full-screen navigation from the List view to the Form panel.

Here's what you'll need to accomplish before you can render your app again:

1. Modify app.js, removing the dependency for the main view and replacing it with the Phone profile.
2. Create the Phone profile file.
3. Generate the common main controller.
4. Build the Phone profile's main controller, extending the common main controller.
5. Create the Phone profile's main view, extending the common main view.

The steps detail how you're going to proceed. The process seems like a lot of work, but believe us when we say that you'll take a lot of little steps that will amount to one giant leap. This is going to be a long section, so please bear with us. You have a lot of exploring to do along the way.

You'll begin this transition by doing some open-heart surgery on app.js. This means that you won't be able to see anything render until you're done, so tread carefully.

The first thing to do is open app.js in your editor and find the following line:

```
views : ['Main'],
```

Replace it with this:

```
profiles : [
    'Phone'
],
```

You're replacing views with profiles because you're going to put the device profiles in charge of requiring the phone-specific main view controller. As you'll see later, it's going to be the controller's responsibility to require the main view and because you're going to be using the Phone profile the phone-specific main view will be loaded and initialized. It's worth noting that we'll revisit this section to add the Tablet profile once you're done with the Phone profile.

The next step to getting the Phone profile setup is to create the JavaScript file app/profiles/Phone.js, as shown in the following listing.

Listing 11.7 Your Phone profile class

```
Ext.define('App.profile.Phone', {          // 1 Defines Phone profile class
    extend: 'Ext.app.Profile',

    config: {
        controllers:[                      // 2 Configures Phone profile controllers
            'Main'
        ],
        views : [                          // 3 Includes Phone profile views
            'Main'
        ]
    },

    isActive: function() {                 // 4 Makes this profile active
        return true; // remove when testing on devices
        return Ext.os.is('Phone');
    },

    launch: function() {
        Ext.create('App.view.phone.Main'); // 5 Renders Phone profile main view
    }
});
```

There may not seem like a lot's going on in listing 11.7, but this code does quite a bit from the application bootstrap perspective. You begin by defining the Phone profile class ❶ which extends `Ext.app.Profile`. Whenever you're going to create a device profile you will need to extend `app.Profile`.

Listing 11.7 has a config section that features a controllers array with `Main` as a single controller ❷. Placing controllers here will tell Sencha Touch to look in app/controllers/phone for the `Main` controller class. If you were to place the main controller in the application definition (which is totally legal), the Phone profile version of the main controller wouldn't be loaded and key features for the phone version of the

> **Profile naming convention**
>
> For this simple application you're creating two simply named profiles, Phone and Tablet. The important thing to note is that even though we're beginning the profile's class Constructor with an uppercase letter (you're following naming standards) classes for each profile will be located using the "package" or "property" naming convention. That means that controller/phone and view/phone will be used for the Phone profile that you've just created.

app wouldn't be properly configured. This could lead to stopping the application from rendering or exceptions being thrown by the framework.

> **NOTE** Because profiles act very much like a subset of an application instance they can load stores, models, views, and controllers. This extremely powerful feature allows you to customize your application down to what features are available based on what profile is active.

You add a `views` directive, which will require the main view from the `App.view.phone` namespace ❸. By requiring the view as part of this profile you're ensuring that the class is loaded in its designated namespace by the time the profile's `launch` method is called.

The `isActive` method ❹ is an important method to discuss because it's where the framework figures out what profile is the correct one to use to render your application. Here you can inject all sorts of tests, such as looking at screen resolution or any other type of attribute that's specific to the device for which you're developing a profile.

In your simple `isActive` method, you've put in a stub that returns true. Doing so ensures that the framework will select this profile even if you test in the desktop browser, where you do 90% of your development tasks. When you're done developing this profile you'll remove that line and have this method return the result of the `Ext.os.is('Phone')` test.

The last method is the `launch` method ❺ which creates an instance of the Phone profile–specific main view, concluding the bootstrapping of the application from the profile's perspective.

With the `Profile` class complete, let's move on to creating the common main controller and its Phone profile extension.

11.5.7 Introducing the common controller

The next listing defines the common `Controller` class. After you've created the ContactsList and ContactDetails views we'll revisit this class to add in the interaction modeling.

Listing 11.8 The common main controller stub

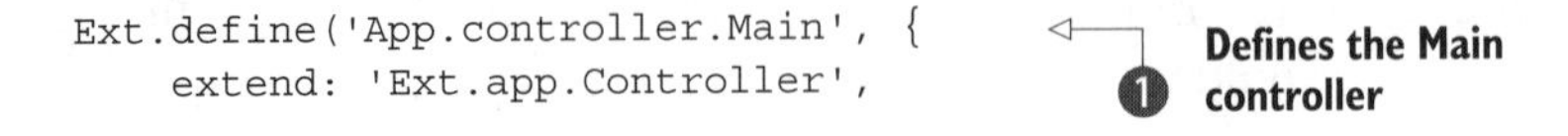

```
Ext.define('App.controller.Main', {
    extend: 'Ext.app.Controller',
```

❶ Defines the Main controller

```
    config : {
        refs : {          ❷ Creates a refs stub

        },
        control : {       ❸ Adds a stub control construct

        },
        views : [         ❹ Includes a views stub

        ]
    }
});
```

Listing 11.8 begins by defining the common main controller ❶. When creating new controller definitions that aren't profile-specific you must extend `app.Controller`. Next you have a stub for instance-based references, known as `refs` ❷, and one for ComponentQuery-based control constructs ❸.

As you'll see later, `refs` will act as a means to set up static references to instances of views. Because this application is relatively simple you'll have references for the main view, contact list, and the details form. As you develop each view you'll circle around and add `refs` as needed.

The control object describes to the framework which views to be on the lookout for and which events from those views to listen to. In addition to the events you describe what methods to call in reaction to the described events. In this case you want to know when a contact is selected in the contact list, when the back button is pressed (Phone profile only), and when the save button is pressed (Tablet profile only). For now, leave it as an empty object stub as a reminder and you'll return to it when you've developed those views and are ready to wire up the event model.

The last bit is an empty views declaration ❹. You added this stub for views that'll be common to both the tablet and phone versions of the app, ContactsList and ContactDetails. This means that as you work to build these views and wire them in you'll revisit this common controller.

Your common main controller may not seem as if it's doing a lot right now. Don't worry; you're still in the beginning phase of developing this application. As you start to develop the views you'll see how it'll play a key role in housing reusable code between the Tablet and Phone profiles.

The next step is to create the Phone profile-specific controller.

11.5.8 Adding the Phone profile controller

You just saw the addition of the common main controller class. As you'll see in the following listing, the Phone profile controller is responsible for phone-specific layout and view management.

Listing 11.9 The Phone profile controller

```
Ext.define('App.controller.phone.Main', {
    extend: 'App.controller.Main',
```

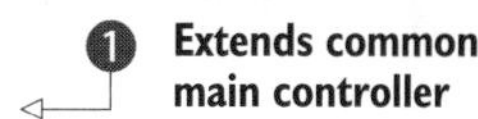

```
    config : {
        anims : {                                          ❷ Sets up reusable
            next : {                                          animation objects
                type      : 'slide',
                direction : 'left'
            },
            previous :  {
                type      : 'slide',
                direction : 'right'
            }
        }
    },
                                                           ❸ Adds event
    showContactDetails: function(record) {}                  listener stub
});
```

To define the Phone profile controller you must extend the common controller you made earlier ❶. Doing so allows you to inherit everything that you'll need for this profile-specific controller to do its job.

Think back to how this application is supposed to be displayed: full screen. This means that when a user taps on an item on the list you must replace the existing List view and display the detailed Form panel. The user can navigate back to the list via the back button. To configure these animations you set up a custom `anims` object ❷, which contains the animation definitions for forward and backward navigation. You set this up in the spirit of reusability. Technically speaking, object creation is expensive. So if you can create once and reuse many times you win and the user wins. These animation objects demonstrate how you can set up reusable objects. You'll implement these objects via getter methods in the `showContactDetails` method ❸.

You have two methods in this class: `showContactDetails` and `onMainBackButton`. Both of them are going to be executed by your common main controller. Because you don't have views to manage you're not going to add code to them yet. `showContactDetails` will be called when a user selects an item in the list, and it'll be responsible for creating an instance of `ContactDetails` (if it doesn't yet exist) and animating it into the viewable space.

With the Phone profile controller now in place you can add the Phone profile main view, which prepares you for the development of the ContactsList and ContactDetails views and wiring up the interaction models between them.

11.5.9 Adding the Phone profile main view

You just set up the Phone profile class and its related controller, paving the way for you to create the Phone profile view. This is where you start to pick up the pace a little; the pieces to this application are starting to come together for the phone version.

Create the following file in your project's filesystem: app/view/phone/Main.js. (See the following listing.)

Listing 11.10 The Phone profile main view

```
Ext.define('App.view.phone.Main', {                    ❶ Defines Phone profile main view
    extend    : 'App.view.Main',

    requires : [
        'Ext.layout.Card'
    ],

    config : {

        layout : 'card',
                                                        ❷ Listens for tap event
        control : {
            'button[ui=back]' : {
                tap : 'onMainBackButtonTap'
            }
        }
    },

    initialize : function() {
        var me = this;
        me.add({                                        ❸ Adds ContactsList view
            xtype : 'contacts'
        });

        me.down('toolbar').add({
            xtype  : 'button',
            text   : 'Back',
            ui     : 'back',
            hidden : true
        });

        me.callParent();
    },
                                                        ❹ Handles back button tap
    onMainBackButtonTap : function(btn) {
        this.fireEvent('back', this, btn);
    }
});
```

A lot's going on in listing 11.10, beginning with the registration of the `App.view.phone.Main` class ❶. Notice how this view extends the common `App.view.Main` class you created earlier. If you compare it to the common main view you can see that it adds to it in a few dimensions.

The first is the requirement of the card layout class and its implementation. Your common main view has no layout required or implemented. Later you'll compare the Phone main view to the Tablet main view, and the Tablet main view will implement an `Hbox` layout to provide a completely different user experience for this application.

In the config object you'll find a control object ❷. This object allows your main view, a subclass of `Container`, to listen for events from a child anywhere in its parent-child hierarchy using ComponentQuery syntax. In this case you're listening to the tap event of the back button that you inject in the `initialize` method ❸. When the tap event is fired the `onMainBackButtonTap` method ❹ is called, allowing you to fire a custom event.

Why are you firing this custom event and not doing any work?

The reason you're firing a custom event is because we want to show you the capability of firing custom events for your views and how to capture them in your controllers. In the `onMainBackButton` method you could do some work such as validating the form values and preventing backward navigation if something is missing or incomplete.

As you'll see later, the control construct you're setting up in this view works the same way as the control constructs that you define in controllers. The reason you define control constructs in views is because you're treating this profile-specific view as a class in its own right, meaning that you can have class-specific events. In this case you fire the `back` event when the back button is pressed, passing along key references for the main controllers (common and profile-specific) to manage navigation.

Control event registration and execution scope

When defining control constructs inside the prototypal config object you use string representation of the method names to be executed by the event system. You do this because the `this` keyword refers to `window`, not the instance of `App.view.phone.Main` when the control object is being parsed by JavaScript (when the class is defined). String representation of method names affords you the flexibility to set up these event listener constructs without the `this` limitation. Event handlers defined via string representation for the control constructs are always called within the scope for the class in which the control construct is being defined.

Inside the `initialize` method you do quite a bit of work. The first step is to add the soon-to-be-created ContactsList view as a child. It doesn't yet exist, so attempting to view our app in the browser at this point will cause an exception to be thrown, rendering the app in a failed state.

You also add a hidden back button to the top-docked toolbar. It's hidden because when the application initializes the ContactsList is the first view to be rendered. The back button doesn't need to be visible until the ContactDetails view is focused. Soon you'll be managing its appearance via updates to the Phone profile's main controller.

The next step in your journey is to build the ContactsList view.

11.5.10 Building the data-driven ContactsList view

The ContactsList view is the first data-driven view in your application and is easy to develop, as shown in the following listing.

Listing 11.11 The ContactsList view

```
Ext.define('App.view.ContactsList', {
    extend : 'Ext.dataview.List',
    xtype  : 'contacts',
```

```
    requires : [ 'App.store.Contacts' ],

    config : {
        itemTpl : '{firstname} {lastname}',
        store   : {
            type : 'contacts'
        }
    }
});
```

There's not much to the ContactsList view, but there are some key points we need to discuss. First is the requirement of the Contacts store class, where you're seeing the documented inclusion of this class for the first time. The Contacts store is implemented via a shortcut object using its alias.

The next step to getting this to work is to ensure that this class is documented as a requirement. The best place to do this is the common main controller because it's a shared view between the Phone and Tablet profiles. To do this you'll have to open the common main controller class (`app/controller/Main.js`) and add the following to the config object:

```
        views : [
            'ContactsList',
//            'ContactDetails'
        ]
```

By adding this snippet into the config object of the common main controller you're ensuring that when that controller is defined your ContactsList is injected into the namespace as a dependency to this controller. The ContactDetails view is commented out because you haven't created it yet and we want you to see part of the phone version of this app in action (figure 11.13).

If you look at the app you can see the ContactsList view for the first time and that the back button in the top toolbar is hidden. Recall that the back button is hidden because you configured it that way when you created the Phone profile's main view class.

At this point you've done a lot of work to get the Phone profile working and you can begin to see the fruits of your labor. But you're not done yet. To finish this profile you'll need to create the ContactDetails view and then wire up the interaction models with the controllers. Afterward you'll create the tablet view and then move on to package this application for production.

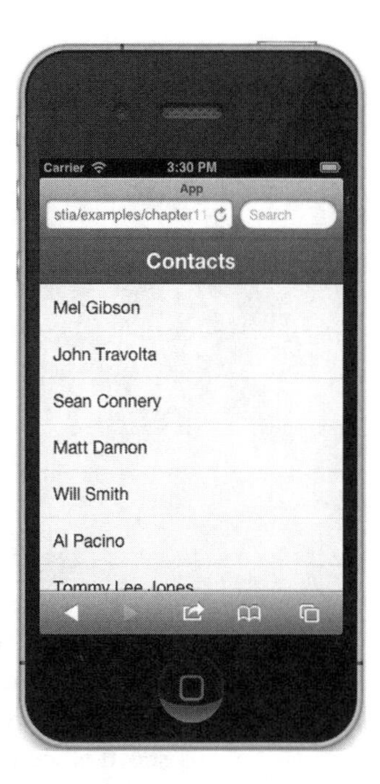

Figure 11.13 Seeing the ContactsList for the first time

11.5.11 Creating the ContactDetails view

The ContactDetails view will be responsible for displaying the details of the contact model selected via the `itemtap` event from the ContactsList and will allow users to modify the data because it extends `form.Panel`. The next listing shows you how to do it.

Listing 11.12 The ContactsDetails view

```
Ext.define('App.view.ContactDetails', {                  <-- ❶ Defines class
    extend : 'Ext.form.Panel',
    xtype  : 'contactdetails',

    requires : [
        'Ext.form.FieldSet'
    ],

    config : {
        items : {
            xtype : 'fieldset',
            title : 'Contact details',

            defaults : {
                xtype          : 'textfield',
                labelWidth     : '30%',
                autoCapitalize : true
            },

            items : [                                    <-- ❷ Adds fields
                {
                    name  : 'firstname',
                    label : 'First'
                },
                {
                    name  : 'lastname',
                    label : 'Last'
                },
                {
                    name  : 'phone',
                    label : 'Phone'
                }
            ]
        }
    }
});
```

The ContactDetails view is simple even though listing 11.12 is rather long. You define the class ❶, extending `form.Panel`, and set an `XType` for lazy instantiation. What makes this listing so big is the number of form items ❷.

To see this view render in your application you're going to need to make a few quick changes to the infrastructure classes. Start by modifying the `views` array in the common main controller class (`app/controller/Main.js`). Uncomment out all of the views in the `views` array so that it looks like this:

```
    views : [
        'ContactsList',
        'ContactDetails'
    ]
```

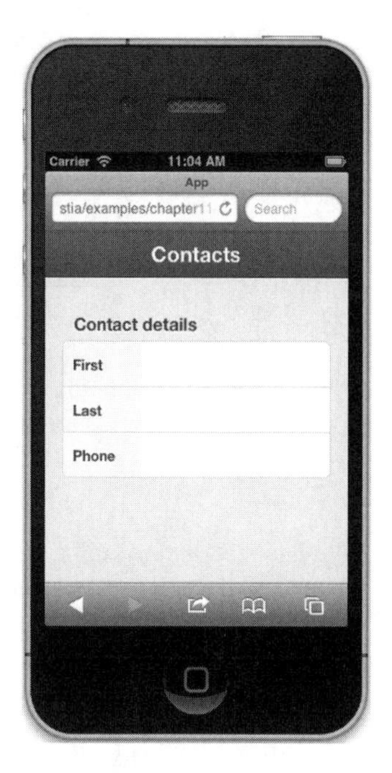

Figure 11.14 Your ContactDetails view rendered on a phone, error-free

By making this change you're instructing the application to load the ContactDetails view as a dependency. This change won't make it viewable yet. You want to see it render so you can detect whether you've made any errors in your code.

The next change is a temporary one, where you'll modify the Phone profile class itself (`app/profile/Phone.js`), forcing this view to render. Modify the `me.add` method call in the `initialize` method so that it looks like this:

```
me.add({
    xtype : 'contactdetails'
});
```

By making this change you're forcing ContactDetails to render instead of the ContactList. This is a temporary change to ensure that you have no problems. Figure 11.14 shows how the ContactDetails view looks on a phone.

You can see that your newly developed view works pretty well. No data is present yet, and that's okay. To get this to work you'll have to wire in the interaction models.

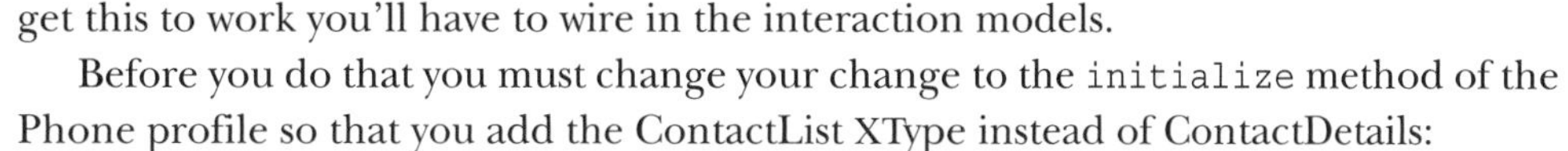

Before you do that you must change your change to the `initialize` method of the Phone profile so that you add the ContactList XType instead of ContactDetails:

```
me.add({
    xtype : 'contacts'
});
```

Let's move on to working on the interaction models.

11.5.12 Wiring up the workflow models into the controllers

Before we dig into the code let's review how this app is going to work. First, users are presented with a list of contacts. They select a contact via a tap event, which will invoke the rendering and displaying of the ContactDetails view via a slide animation. It's at this time that the back button appears at the top-docked toolbar. Users can elect to make changes while on the ContactDetails screen. Whenever the back button is pressed you'll ensure that those changes are persisted in the model and the ContactsList view is animated into view.

You'll begin by modifying the common main controller (`app/controller/Main.js`) to inject some commonly used refs, control constructs, and methods. Instead of giving you a step-by-step instruction guide on how to change the common main controller we'll show you the final class (see the next listing) and dive into it after you've read it.

Listing 11.13 The common main controller updated

```
Ext.define('App.controller.Main', {
    extend: 'Ext.app.Controller',

    config : {
        refs : {                                        <-- 1 Configures static references
            mainView    : 'main',
            contacts    : 'contacts',
            details     : 'contactdetails'
        },

        control : {                                     <-- 2 Sets up event listener
            contacts : {
                select : 'onContactSelect'
            }
        },

        views : [
            'ContactsList',
            'ContactDetails'
        ]
    },

    onContactSelect: function(list, record) {           <-- 3 Handles ContactList select
        this.showContactDetails(record);
    },

    persistContact : function() {                       <-- 4 Persists data in model
        var details = this.getDetails(),
            record  = details.getRecord();

        if (! record) {
            return;
        }
        record.beginEdit();
        record.set(details.getValues());
        record.endEdit();
        record.commit();
    },

    /**
     * Override for each profile
     */
    showContactDetails  : Ext.emptyFn,                  <-- 5 Implements showContactDetails
    onMainBackButton    : Ext.emptyFn
});
```

The updates to your common main controller begin with the addition of the static references via the `refs` key ❶. These references work by means of key-value pairs to tell the controller how to find an instance of a component via ComponentQuery. We should spend a moment discussing this further because it's a key feature that can be useful.

Let's take the first ref, `mainView`. You see the word `mainView` as the key and `main` as the value of this object. The value (the right side of the colon) is the ComponentQuery syntax, where you're looking for any class with the `XType` set to `main`. The key (the left side of the colon) is set to the same name and will be used to automatically create the reference and getter method. Later you'll use the self-generated `getMainView` method in the Phone profile controller to get a reference to the class that the ComponentQuery found.

> **Refs designed for single instances**
> Lots of people get confused by refs when they're trying to access an instance of a class where many instances may exist at any one time. It's helpful to clear up this confusion early by highlighting the fact that refs are designed for single instances only. When developing apps we typically use them as pointers to the main view. All other views typically can be thrown away, but the main view generally stays there to manage navigation for the entire application. So the rule of thumb is that if you need to access more than one instance of a view, don't use refs. We'll show you how to leverage the control constructs to manage view instance awareness via event handlers.

You have the control constructs ❷ to set up a select listener for the `ContactsList` instance where the local member `onContactListSelect` will be executed. There's also a back event listener for the main view instance that'll trigger the execution of `onMainBackButton`.

The `onContactSelect` method ❸ is defined as a means to call the `showContactDetails` ❺ member, passing in only what's required, which is the record. `showContactDetails` is a method that has different implementations for each profile, so because you're in the common main controller you create empty functions that are marked via comments to be overridden by profile-specific controller subclasses.

`persistContact` ❹ is a method that's defined to merge the data from the ContactDetails screen (form) to the list. It's essentially a method that'll be called in two use cases. The first is when the user is viewing a contact on the Phone profile and hits the back button. The second is when the person is viewing a contact on the Tablet profile and hits the save button.

> **Why no save button in the Phone profile?**
> Without getting into too much user experience theory, in most apps changes are persisted when users modify data and navigate away from the screen that they made those changes in. Not only does this eliminate the need for the screen real estate to display the save button, but it allows the app to be used in a manner where users "get in and get out," which is the goal for the best-designed phone applications. For tablet apps you generally have the screen real estate, so you'll add a save button to the ContactDetails form when you build that profile and controller workflow model.

You're almost ready to wrap up the Phone profile. To do so you need to get the Phone profile main controller wired up with the application. It'll implement the `showContactDetails` method and add another.

11.5.13 Adding the Phone profile main controller

As you'll see, the Phone profile main controller (app/controller/phone/Main.js) is going to extend the common main controller and implement the showContactDetails and onMainBackButton methods, effectively achieving forward and backward navigation between the two screens.

The next listing is long, but you've seen a third of it. The last two-thirds are the two method implementations and are new, so we'll put our focus there.

Listing 11.14 Phone profile main controller

```
Ext.define('App.controller.phone.Main', {
    extend: 'App.controller.Main',

    config : {
        views : ['App.view.phone.Main'],
        anims : {
            next : {
                type      : 'slide',
                direction : 'left'
            },
            previous :  {
                type      : 'slide',
                direction : 'right'
            }
        },
        control :{                                         // 1 Listens to back event
            main : {
                back : 'onMainBackButton'
            }
        }
    },

    showContactDetails: function(record) {                 // 2 Shows ContactDetails form
        var me              = this,
            mainView        = me.getMainView(),
            contactsList    = me.getContacts(),
            lastName        = record.get('lastname'),
            firstName       = record.get('firstname'),
            contactDetails = mainView.down('contactdetails'),
            nextAnim        = me.getAnims().next;

        if (! contactDetails) {                            // 3 Creates ContactDetails view once
            contactDetails = mainView.add({
                xtype      : 'contactdetails',
                hideFields : false
            });
        }

        contactDetails.setRecord(record);

        mainView.setTitle(firstName + ' ' + lastName);
        mainView.down('[ui=back]').show();
        mainView.animateActiveItem(contactDetails, nextAnim);

        Ext.defer(contactsList.deselectAll, 1, contactsList);
    },
```

```
    onMainBackButton : function() {
        var me       = this,
            main     = me.getMainView(),
            contacts = main.down('contacts');

        me.persistContact();

        main.setTitle('Contacts');
        main.down('[ui=back]').hide();
        main.animateActiveItem(contacts, me.getAnims().previous);
    }
});
```

❹ **Handles back nav, persists data**

The additions to the Phone profile controller in listing 11.14 may seem like a lot at first glance, but they're relatively minor. Here's the gist.

The first thing we want you to focus in on is how you register listeners to the back event ❶ of the main view via ComponentQuery syntax. `onMainBackButton` will be called when the user presses the back button, and it'll be responsible for retrieving the values from the form, merging the values into the selected record from the list, and thus effectively persisting the changes in memory. The method will also be responsible for animating the list back into the viewable space.

The `showContactDetails` method ❷ is called when an item is selected. Recall that the workflow begins with a select event being fired from ContactDetails (see listing 11.14), which causes the common main controller's `onContactSelect` method to execute. `onContactSelect` then executes `showContactDetails`.

Inside this method you gain several references at the very top for later use in the rest of the method. You first hit an `if` condition, where you create an instance of ContactDetails the first time this method is called ❷. Next you instruct the instance of ContactDetails to bind the selected contact record instance via its `setRecord` method, allowing the ContactDetails screen to make the data editable. You also animate the ContactDetails screen into view ❸ via the main view's `animateActiveItem` method using the preconstructed and reusable next animation object set up in the controller's prototypal config block.

When the user hits the back button on the application it causes the tap event to be fired. The tap event is listened to via the common main controller (listing 11.14), which will cause the `onMainBackButton` method ❹ to be called.

Inside this method you call the inherited `persistContact` method, found in the common main controller, to merge the data from the ContactDetails screen to the current instance of the model.

The last job of this method is to animate the ContactList back into view via the reusable previous animation configuration object defined in the prototypal config object.

This wraps up the Phone profile version of this application. Fire it up on your phone or browser and exercise the workflows. Figure 11.15 shows a quick demo of us working with the app utilizing one full cycle of the workflow.

The application workflow for the Phone profile seems to work well. The next thing you need to do is to look at this application in a tablet (figure 11.16).

Figure 11.15 The full application workflow cycle in action

The first thing you might think is this thing looks silly. And it does! It looks silly because of the gobs of wasted space. The reason you have all of this wasted space is because an HTML5 app that was designed for a phone experience can render on a tablet, but it begs some serious design reconsideration.

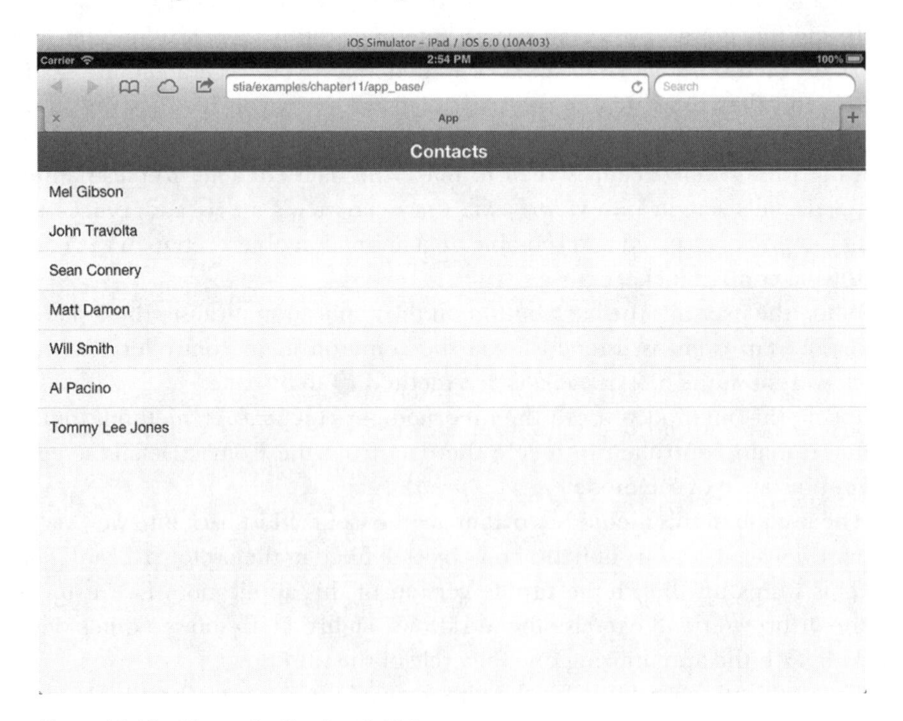

Figure 11.16 The application in a tablet

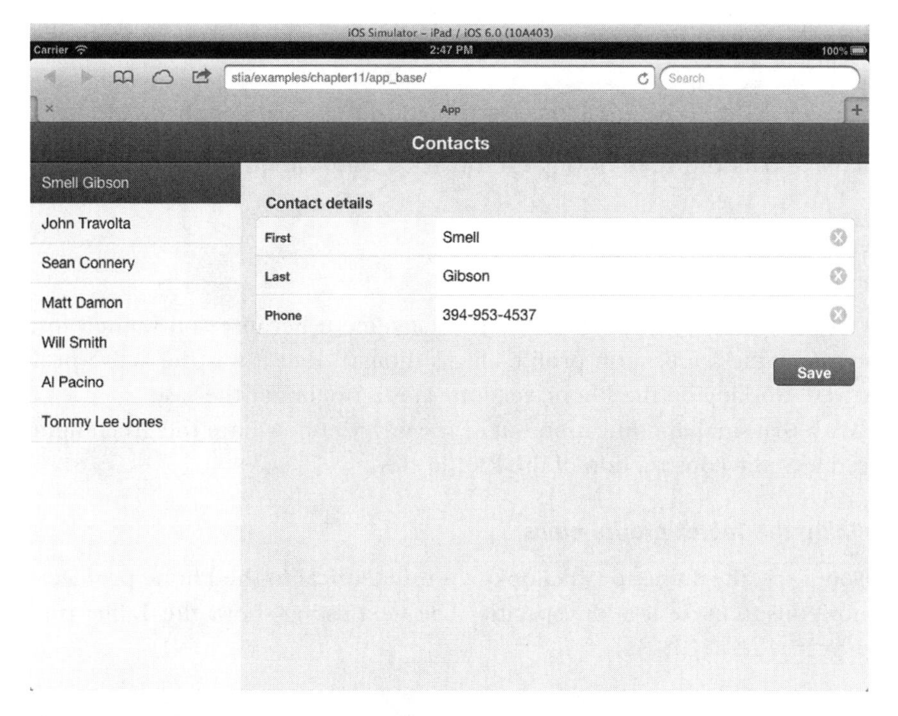

Figure 11.17 The app in a tablet simulator

Given screen-size difference the user interaction model will change a bit. Figure 11.17 shows what your app should look like on the tablet.

As you compare the phone version to the tablet version you can see that you'll need to make changes. First, note that there's no back button in the tablet version. Second, you'll need to inject a save button underneath the form panel.

We hope that you're excited to get close to completing this. Let's switch gears and get the tablet version done. You're going to move at a much faster pace because you've already experienced a lot of what you're about to do.

11.6 Building the tablet version of your application

To develop the Phone profile version of your app you had to do quite a bit! Recall that you worked to set up the infrastructure for this application, which includes the data model and store. The tablet version of your app will require some work but it'll reuse over 98 percent of the existing code.

Here's the full list of items you need to accomplish for the Tablet profile:

- Add the Tablet profile class.
- Construct the view.
- Wire in the interaction models with the tablet controller.

Before you begin working on the Tablet profile you must modify the Phone profile. Currently you have the `isActive` method set up like this:

```
isActive : function() {
    return true;
    return Ext.os.is('Phone');
},
```

You need to modify it, removing the `return true;` statement, so it looks like this:

```
isActive : function() {
  return Ext.os.is('Phone');
},
```

Recall that you added the `return true;` statement because you wanted to force the case where the application profile always thought that it was the active profile when you were working on the Phone version. This is no longer the case.

With that small modification out of the way, you now have the green light to move ahead with the construction of the Profile class.

11.6.1 Building the Tablet profile class

As you'll see, the Tablet profile looks almost identical to the Phone profile, with a few minor edits to make it tablet specific. The next listing shows the Tablet profile class, app/profile/Tablet.js.

Listing 11.15 The Tablet profile class

```
Ext.define('App.profile.Tablet', {
    extend: 'Ext.app.Profile',

    config: {
        controllers:[
            'Main'
        ],
        views : [
            'Main'
        ]
    },

    isActive: function() {                                    ❶ Forces tablet profile to be active
        return true;  // remove before production build
        return Ext.os.is('Tablet') || Ext.os.is('Desktop');
    },

    launch: function() {                                      ❷ Creates main view
        Ext.create('App.view.tablet.Main');
    }
});
```

To construct the Tablet profile class you follow the same pattern that you did with the Phone profile. This includes the immediate `return true;` of the `isActive` method ❶, because you're working on this application at the moment. Just as you did with the Phone profile, you'll remove this `return true;` statement before you move on to working toward the production build.

Very much like the Phone profile, the `launch` method ❷ of the Tablet profile is responsible for creating the instance of the Tablet profile main view subclass that you'll be creating next.

11.6.2 Constructing the tablet main view

To construct the tablet main view you'll follow a similar pattern to the phone main view, with a twist. The following listing contains `app/view/tablet/Main.js`.

Listing 11.16 The tablet main view

```
Ext.define('App.view.tablet.Main', {
    extend : 'App.view.Main',

    requires : [
        'Ext.layout.HBox'
    ],

    config     : {
        layout : {
            type   : 'hbox',
            align  : 'stretch'
        }
    },

    initialize : function() {
        var me = this;
        me.add([
            {
                xtype : 'contacts',
                flex  : 1
            },
            {
                xtype : 'contactdetails',
                flex  : 3
            }
        ]);

        me.callParent();

        me.down('contactdetails').add({
            xtype  : 'button',
            text   : 'Save',
            ui     : 'confirm',
            itemId : 'saveDetails',
            style  : 'float:  right; margin-right: 10px;',
            width  : 100
        });
    }
});
```

❶ Requires HBox layout

❷ Adds custom views

❸ Injects save button

To implement the different layout you first require and then implement the `HBox` layout ❶, giving you the flexibility to have the ContactList and ContactDetails forms side by side. Inside the `initialize` function you're adding the ContactList and ContactDetails views as children ❷ at the same time. After you execute `this.callParent()`, you immediately inject a save button as a child of the ContactDetails view ❸.

Why are you injecting a button into the view from the profile?

The reason you're injecting a button into the view from the profile is because the whole point of the device profiles is to inject changes into views and application behaviors. If you elected to add this logic to the ContactDetails view you'd negate the purpose of the profile, which would make your application much more rigid and difficult to maintain.

With the view complete you can move on to getting the tablet main controller done and then look at how your app behaves.

11.6.3 Constructing the tablet controller

Relative to the phone main controller, the tablet main controller in the following listing is simple. This is because there's no navigation to manage for this application. You simply have to manage the selection of an item and then persist the changes if someone taps on the save button.

Listing 11.17 The tablet controller

```
Ext.define('App.controller.tablet.Main', {
    extend: 'App.controller.Main',

    config : {
        control : {
             'button[itemId=saveDetails]' : {         ❶ Listens to save button tap event
                tap : 'onSaveButtonTap'
             }
        }
    },

    showContactDetails: function(record) {             ❷ Shows data in Form panel
        this.getDetails().setRecord(record);
    },

    onSaveButtonTap : function() {             ❸ Persists data to contact model
        this.persistContact();
    }
});
```

To listen to the save button tap event you must set up a control construct ❶ using ComponentQuery syntax. Here you're searching any button with an `itemId` of `saveDetails`. You could've gotten a lot more specific with that query, but for our purposes this is good enough.

The next thing to look at is the `showContactDetails` method ❷. In the Phone profile main controller you had to worry about creating an instance of ContactDetails, binding the record, and then animating it in. In the tablet main controller all you have to worry about is binding the record to the instance of the ContactDetails screen.

The `onSaveButtonTap` method ❸ is called when the tap event is fired by the newly injected save button (from the Tablet profile view). This method will call `this.persistContact()`. You could inject a bit of work here if you wanted to, such as verifying that the form fields are valid and electing to persist the values if the data is clean and present, or showing a MessageBox alert dialog that tells the user about data issues that they must resolve.

You have one more task to do, and that's to tell your application that you have a new device profile to load. To do so you'll need to open app.js and modify the `profiles` array so it includes the Tablet profile:

```
profiles : [
  'Phone',
  'Tablet'
],
```

Now load up your app in your tablet simulator or browser and see if everything works as advertised (figure 11.18).

You've just seen what it takes to build an application from start to finish. You used Sencha Cmd to generate the application structure and you created the necessary model, store, views, profiles, and controllers to support a tablet and phone version of the application.

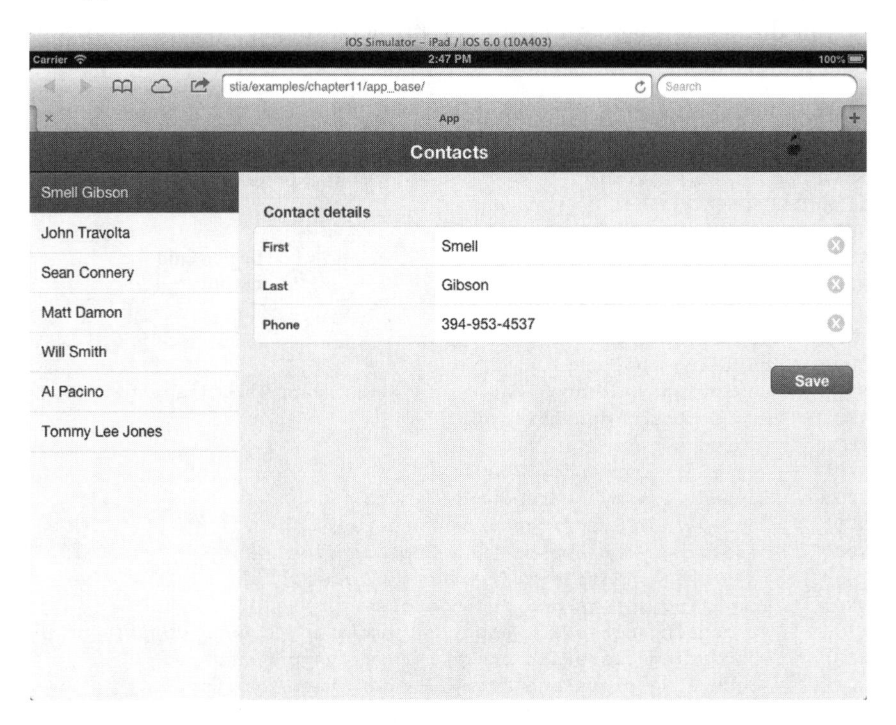

Figure 11.18 The Tablet profile of your application works well on a tablet simulator.

Before we move on to making testing and production builds let's do a little housekeeping. Open the Tablet profile, `app/profile/Tablet.js`. Next, remove the `return true;` statement from the `isActive` method so that it looks like this:

```
        isActive: function() {
        return Ext.os.is('Tablet') || Ext.os.is('Desktop');
    },
```

The last leg of this journey involves generating a testing build and then migrating to production, where we'll introduce you to the Touch Microloader.

11.7 Packaging your application for web deployment

You're currently at step 3 of the app development process. At this point you've deemed the application feature rich enough to begin the development testing phase.

In this section you'll explore HTML5 packaging options. This includes going through the testing and production build phases to ensure your application operates the way you want. You'll begin with the testing build.

11.7.1 Creating a testing build

You're now at step 4 of the development process (generate and verify testing build). To create a testing build you must drop down to a shell and change directories to where your application resides. (We changed directories to where we generated the application, /www/myapp.)

From there you're going to issue the proper Sencha Cmd syntax to generate the testing build. The next listing shows how to do it.

Listing 11.18 Generating a testing build

```
sencha app build testing                    <-(1) Issues proper command
Sencha Cmd v3.0.0.190
[INFO]     init-properties:                 <-(2) Begins Cmd output
[INFO]     init-sencha-command:
[INFO]     init:
[INFO]     -before-app-build:
[INFO]     app-build-impl:
[INFO]     building application
[INFO]     Deploying your application to /www/myapp/build/App/testing
[INFO]     Copied touch/sencha-touch.js
[INFO]     Copied app.js
[INFO]     Copied resources/css/app.css
[INFO]     Copied /www/myapp/resources/icons
[INFO]     Copied /www/myapp/resources/startup
[INFO]     Resolving your application dependencies
                  (file:////www/myapp/index.html)   <-(3) Resolves dependency
[INFO]     Compiling app.js and dependencies
[INFO]     Processing classPath entry : /www/myapp/sencha-compile-temp-dir
[INFO]     Processing classPath entry : /www/myapp/touch/src
[INFO]     Processing classPath entry : /www/myapp/app.js
[INFO]     Processing classPath entry : /www/myapp/app
[INFO]     Processing class inheritance graph
```

```
[INFO]    Processing instantiation refereces to classes and aliases
[INFO]    Processing source dependencies
[INFO]    Concatenating output to file
                   /www/myapp/build/App/testing/app.js          <-- 4 Concatenates JavaScript file
[INFO]    Completed compilation.
[INFO]    Processed local file touch/sencha-touch.js
[INFO]    Processed local file app.js
[INFO]    Generated app.json
[INFO]    Embedded microloader into index.html
[INFO]    -after-app-build:
[INFO]    app-build:
```

Sencha Cmd does quite a bit to generate a testing build. It all begins with the submission of the `sencha app build testing` command ❶, which causes a flurry of activity from the shell console ❷.

As you can tell from the Cmd output, resources are copied to build/App/testing relative to where your application resides. When invoking any type of build Cmd will analyze (not execute) your application's dependency structure ❸ and generate a dependency model, which it uses in conjunction with the app.json file to concatenate all of your JavaScript classes into a single file named app.js ❹.

Remember our conversation about dependency modeling?

Recall our conversation where we were discussing documenting your application's JavaScript requirements via both the `requires` statement in your class definitions and app.json modifications. If you have files that aren't documented properly you might find yourself with a bad testing or production build. The best way to detect these things is to look at the Network tab of the WebKit inspection tool. There you'll likely find that JavaScript files other than app.js are being loaded. In this case you want to revisit your application definition and add those files to the app.json file or the dependent class(es).

Sencha Cmd places app.js in the root of your project build directory (<app_location>/build/App/testings). It's worth noting that app.js contains all of the required Sencha Touch code as well as whitespace and comments; thus it isn't intended for production (generally internet) deployments.

> **NOTE** If you'd like to change where Sencha Cmd places builds you can modify them in the section labeled "buildPaths" in app.json. You can use relative or absolute paths.

After the build process completes you can look at it in the browser via http://yourhost/build/App/testing. Because the application will look the same as the previous listings there's no point showing you how the app renders. Focus your attention on the Network tab of the WebKit inspector tool (figure 11.19). As you can see, you have few resources being requested and loaded relative to the development mode of your application. This has to do with the fact that Sencha Cmd has concatenated your files into a single app.js.

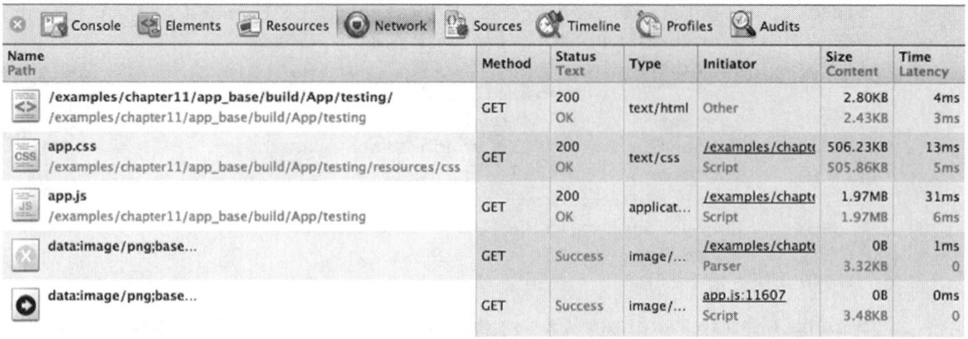

Console Elements Resources Network Sources Timeline Profiles Audits

Name Path	Method	Status Text	Type	Initiator	Size Content	Time Latency
/examples/chapter11/app_base/build/App/testing/ /examples/chapter11/app_base/build/App/testing	GET	200 OK	text/html	Other	2.80KB 2.43KB	4ms 3ms
app.css /examples/chapter11/app_base/build/App/testing/resources/css	GET	200 OK	text/css	/examples/chapte Script	506.23KB 505.86KB	13ms 5ms
app.js /examples/chapter11/app_base/build/App/testing	GET	200 OK	applicat...	/examples/chapte Script	1.97MB 1.97MB	31ms 6ms
data:image/png;base...	GET	Success	image/...	/examples/chapte Parser	0B 3.32KB	1ms 0
data:image/png;base...	GET	Success	image/...	app.js:11607 Script	0B 3.48KB	0ms 0

Figure 11.19 The resources loading from a testing build

> **A quick word about the testing version bootstrap**
>
> We went over, in pretty good detail, how the Microloader application bootstrap works in the development version. The reason we're not covering the testing build Microloader bootstrap is because it's nearly identical to that of the development version. The only major difference between the development and production versions of the Microloader that you need to be aware of is that the Microloader for development is loaded via an external script tag request, whereas the testing version of the Microloader is embedded inside index.html.

Your testing build can now officially be used for user acceptance testing (UAT). Step 5 of the app development process dictates that you should verify whether it's okay to produce a production build. Because you've tested your application successfully, you can move on to step 6: producing and verifying a production build.

11.7.2 *Creating a production build*

The creation of a production build is similar to that of a testing build. The next listing shows what one looks like.

Listing 11.19 Generating a production build

```
sencha app build production                    <-- 1 Generates production build
Sencha Cmd v3.0.0.190
[INFO]  init-properties:
[INFO]  init-sencha-command:
[INFO]  init:
[INFO]  -before-app-build:
[INFO]  app-build-impl:
[INFO]  building application
[INFO]  Deploying your application to /www/myapp/build/App/production
[INFO]  Copied app.js
[INFO]  Copied resources/css/app.css
[INFO]  Copied /www/myapp/resources/icons
[INFO]  Copied /www/myapp/resources/startup
```

```
[INFO]  Resolving your application dependencies                     <--  ❷ Resolves application
                                (file:////www/myapp/index.html)           dependencies
[INFO]  Compiling app.js and dependencies
[INFO]  Processing classPath entry : /www/myapp/sencha-compile-temp-dir
[INFO]  Processing classPath entry : /www/myapp/touch/src
[INFO]  Processing classPath entry : /www/myapp/app.js
[INFO]  Processing classPath entry : /www/myapp/app                  Concatenates ❸
[INFO]  Processing class inheritance graph                           dependency
[INFO]  Processing instantiation refereces to classes and aliases
[INFO]  Processing source dependencies
[INFO]  Concatenating output to file                                  <--
                          /www/myapp/build/App/production/app.js
[INFO]  Completed compilation.
[INFO]  Processed remote file touch/sencha-touch.js
[INFO]  Processed local file app.js                               ❹ Minifies
[INFO]  Minified app.js                                         <--   resources
[INFO]  Minified resources/css/app.css
[INFO]  Generated app.json                                    ❺ Injects production
[INFO]  Embedded microloader into index.html              <--    Microloader
[INFO]  Generating appcache
[INFO]  Generating checksum for appCache item: index.html    <--  Creates
[INFO]  -after-app-build:                                          checksum
[INFO]  app-build:                                               ❻ for build
```

If you scan the output of a production build you'll find that it's close to the testing build process. It all begins with the submission of the `sencha app build production` command ❶ in the shell. Just like in the testing build, application dependencies are resolved ❷ and those dependencies are concatenated into a single file called app.js ❸. This is where the similarities end for the most part.

After the file concatenation phase app.js is minified ❹. This is where the file size is reduced by the removal of unnecessary whitespace and comments. Also the code is modified so that its footprint is smaller (obfuscation).

The last few important items have to do with the production version of the Microloader ❺. After a production version of app.json is created index.html is created with the production version of the Microloader, effectively arming it with a caching mechanism that we'll explore in a little bit. In order for the Microloader to work properly a checksum has to be generated ❻ for this production build.

You can view the production build of your application by visiting http://yourhost/build/App/production/ in your browser. This sets the stage for us to discuss the Microloader.

11.7.3 A deep dive into the production Microloader

As we've discussed in the past, the Sencha Touch Microloader is responsible for bootstrapping and loading necessary resources to kick off the application in your browser. The production version of the Microloader adds features to your application that allow your apps to bootstrap lightning-fast after the first launch.

To put our conversation into context let's peer into what resources are loaded when the production version of your application launches for the first time (figure 11.20).

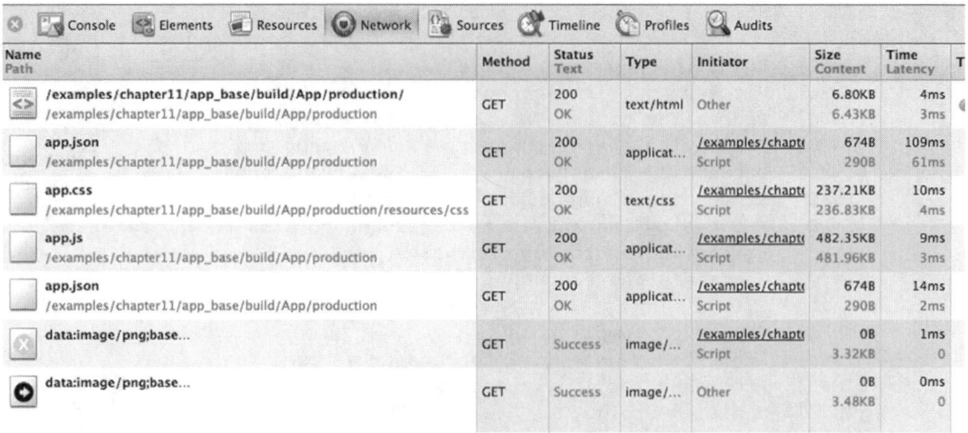

Console Elements Resources Network Sources Timeline Profiles Audits

Name Path	Method	Status Text	Type	Initiator	Size Content	Time Latency
/examples/chapter11/app_base/build/App/production/ /examples/chapter11/app_base/build/App/production	GET	200 OK	text/html	Other	6.80KB 6.43KB	4ms 3ms
app.json /examples/chapter11/app_base/build/App/production	GET	200 OK	applicat...	/examples/chapt(Script	674B 290B	109ms 61ms
app.css /examples/chapter11/app_base/build/App/production/resources/css	GET	200 OK	text/css	/examples/chapt(Script	237.21KB 236.83KB	10ms 4ms
app.js /examples/chapter11/app_base/build/App/production	GET	200 OK	applicat...	/examples/chapt(Script	482.35KB 481.96KB	9ms 3ms
app.json /examples/chapter11/app_base/build/App/production	GET	200 OK	applicat...	/examples/chapt(Script	674B 290B	14ms 2ms
data:image/png;base...	GET	Success	image/...	/examples/chapt(Script	0B 3.32KB	1ms 0
data:image/png;base...	GET	Success	image/...	Other	0B 3.48KB	0ms 0

Figure 11.20 The application resources loaded in the production build for the first time

> **Production Microloader internals**
>
> You've learned how the development (and testing) modes work. The internals of the production version of the Microloader are somewhat complex, so we won't be covering how it works. We'll highlight what you need to know to ensure that your application deployments are successful.

Note the loading of app.json, app.css, and app.js. These three files are required to bootstrap this application. app.json contains the necessary data for the production Microloader to know what build of the application is being loaded. There's more to the relationship between app.json and the Microloader that we'll be revisiting shortly.

> **NOTE** Much like the testing build, the production Microloader is embedded inside index.html. The difference between the two is that the production version is minified; thus it's difficult to read when looking at index.html within your editor.

For now, let's see what happens when the application is refreshed in the browser. Figure 11.21 shows the WebKit inspector tool's Network tab with the same application loaded a second time.

When comparing the resources loaded in an initial load of a Sencha Touch production app (figure 11.20) and successive loads (figure 11.21) you'll find a stark difference. There are no requests for app.css and app.js.

How could this be? How does your application work without those resources being requested from the server?

The answer is quite simple. The production Sencha Touch Microloader caches your application JavaScript and CSS inside HTML5 offline storage, known as local storage. Here's how this all works.

Console Elements Resources Network Sources Timeline Profiles Audits

Name Path	Method	Status Text	Type	Initiator	Size Content	Time Latency
/examples/chapter11/app_base/build/App/production/ /examples/chapter11/app_base/build/App/production	GET	200 OK	text/html	Other	(from c...	1ms 1ms
app.json /examples/chapter11/app_base/build/App/production	GET	200 OK	applicat...	/examples/chapte Script	674B 290B	14ms 2ms
data:image/png;base...	GET	Success	image/...	/examples/chapte Parser	0B 3.32KB	1ms 0
data:image/png;base...	GET	Success	image/...	Other	(from c...	Pending

Figure 11.21 The application resources loaded in the production build for the second time

> **NOTE** If this is the first time that you've gotten exposure to local storage you can learn more by visiting www.html5rocks.com/en/features/storage.

When your application launches for the first time (figure 11.20) the Microloader kicks off an Ajax request for app.json. From here the Microloader decides what to do next after it knows more about your application.

To further the conversation, let's look at the (formatted) contents of app.json:

```
{
    "id": "c64fe431-3558-435e-9b52-40f3f7f088af",
    "js": [
        {
            "update": "delta",
            "path": "app.js",
            "type": "js",
            "bundle": true,
            "version": "8d652b33d85f3ef0a80b47f9064d88474fd020ff"
        }
    ],
    "css": [
        {
            "update": "delta",
            "path": "resources/css/app.css",
            "type": "css",
            "version": "addd145186967b17aaf5440d171933031f33f2d8"
        }
    ]
}
```

In app.json the application ID is listed as well as data items for your JavaScript file and CSS file. Each file has its own version, which is a checksum of that current build. These are vital data points for the production Microloader to perform the following decision.

During an initial load of your application the production Microloader checks local storage to see if your application resources are there. If they aren't it requests them (figure 11.20) from your web server and caches them in local storage. For subsequent loads it loads app.json and verifies that the application ID and file checksum match what's in local storage.

```
> localStorage
<· ▼Storage
    c64fe431-3558-435e-9b52-40f3f7f088af-http://touch/app/app.js: "/*8d652b33d85f3ef0a80b
    c64fe431-3558-435e-9b52-40f3f7f088af-http://touch/app/app.json: "{"id":"c64fe431-3558
    c64fe431-3558-435e-9b52-40f3f7f088af-http://touch/app/resources/css/app.css: "/*addd1
    length: 3
  ▶__proto__: Storage
```

Figure 11.22 The application JavaScript and CSS cached inside local storage

Figure 11.22 shows what things look like in local storage for our production build. If you want to look at it in your app, jump to the JavaScript console, type in `localStorage`, and press Enter. You'll see something similar to figure 11.22.

What you see is that the production Microloader stores your code based on the application ID. Each file is marked with the checksum. If you were to make changes to this application you'd see that the Microloader reacts differently.

Let's walk through a simple change and rebuild things in production. You'll be moving pretty quickly through this.

11.7.4 *Creating a production delta build*

To create a delta production build you have to modify some code. Rather than making a functional change you'll make a cosmetic one.

Open app/view/Main.js in your editor and change the top toolbar title from

```
title : 'Contacts'
```

to

```
title : 'My Contacts'
```

This will give you the platform to generate a production delta build. Make sure that you didn't mess anything up by looking at your application inside the development version (figure 11.23).

You have no syntactical errors in the development version of your application which means that you can go ahead with a testing build. Drop to a command shell and enter the following:

```
sencha app build testing
```

After the build process completes you can look at it in the browser at http://yourhost/build/App/testing. Did you see any exceptions? No? Do you see the change? Yes? Awesome!

Next, let's generate a production build. At the command line, enter

```
sencha app build production
```

For this and subsequent production builds you're going to see output where Sencha Cmd recognizes that there has been a change in the codebase and will generate one or more delta files.

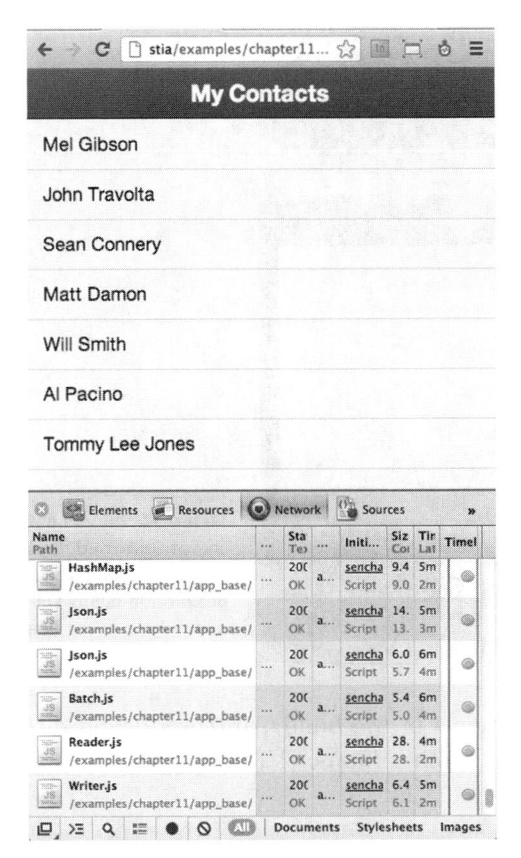

Figure 11.23 Your application change in development mode

Here's what one of those info lines looks like for your application (it's rather large!):

```
[INFO ]Generated delta for: app.js from hash:
 '8d652b33d85f3ef0a80b47f9064d88474fd020ff' to hash:
 '1c30d0bbd2cd15e8f05ffda8a94328b6a20939fe'
```

What you're seeing is that Microloader has noticed that the concatenated build file, app.js, is different, and a different version has been laid down into your production build. This causes app.json to be modified with the relevant changes, which will cause the Microloader to react on the next launch of your application.

From here you can load your application in a browser and see the `onUpdated` method called from the Microloader (figure 11.24).

Because the production Microloader found a delta in app.json it'll call the `onUpdated` method of your application instance, causing the message box (figure 11.24, on the left) to be displayed, giving users the opportunity to load the change now or defer it until the next launch of the application. By this time the application delta has already been stored into local storage, so the user will see the change immediately on the next load of this application.

Figure 11.24 By default (per the `onUpdated` method defined in app.js) your production builds will notify users if there has been an update to the application code.

From here you can deploy the contents of the build/App/production directory to your production web server and serve your customers! This wraps up the application development process!

11.7.5 Where to go from here?

By all standards, the application you developed was simple. The reason it was so simple is because we wanted to focus on the semantics of application development including MVC and Sencha Cmd and everything in between.

There are lots of example applications in the wild that demonstrate all types of design patterns, including DiscoverMusic, an application that this author (Jay) co-authored, which demonstrates how to create custom navigation and views with Sencha Touch 2, providing a cool tablet experience (figure 11.25).

To see this application in action point your modern 10-inch tablet's browser to the following URL: http://discovermusic.senchafy.com/. This application is completely open source and you can get it at the following GitHub URL: https://github.com/ModusCreateOrg/DiscoverMusic.

You can find other applications in the Sencha Touch application gallery at www.sencha.com/apps/. Sometimes authors will provide URLs to demos of their applications and, if their project is open source, the URL for the source code.

Lastly, the Sencha community forum at http://sencha.com/forum/ is an excellent place where people post show-and-tell articles of their applications, contributing to a ton of resources to learn more about developing applications with Sencha Touch 2.0.

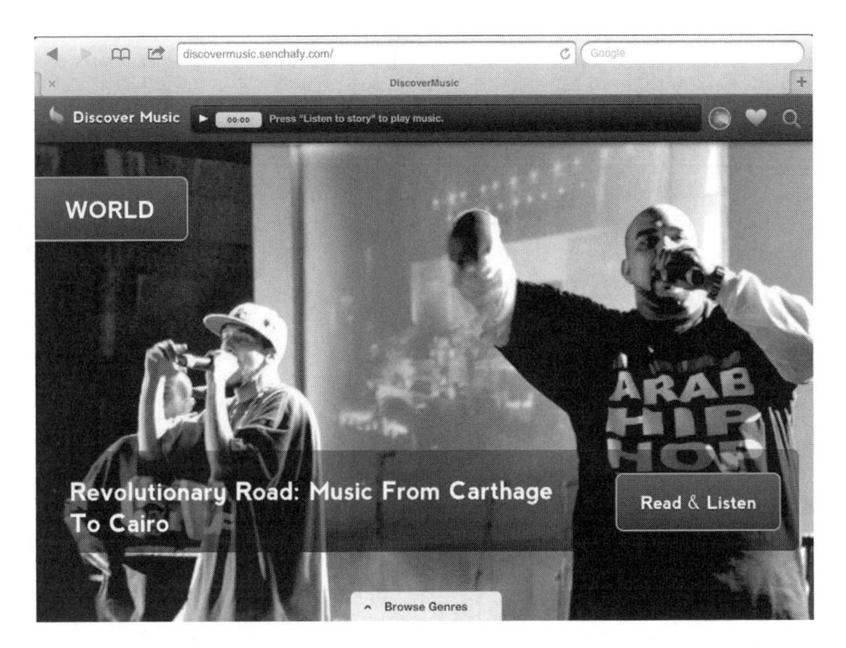

Figure 11.25 DiscoverMusic is an excellent resource for learning how to build custom applications with Sencha Touch 2.0.

11.8 Summary

Wow! You've made it! This has been an incredible journey. You've learned a ton in this final chapter.

You started by taking a quick look at a basic development workflow process that begins with creating a simple application with Sencha Cmd, all the way to the deployment of a production application. The application development kicked off with a deep inspection of what this application does and a walk through the namespace. You started the code with the creation of an application project via Sencha Cmd. From there you inspected Sencha Cmd and the files laid down by Cmd, the foundation for your application.

We walked through the development of a basic application. You used that simple application to learn about Sencha Touch's MVC. You learned key things, from how to leverage control constructs for custom views to how to fire custom events for your application. You also saw how to implement basic profiles for tablets and phones.

After developing the application you saw what it was like to leverage Sencha Cmd to create testing and production builds for web deployments. During this phase you learned how the production Microloader accelerates application bootstraps by use of the Microloader.

index

A

E

F

G

H

I

U

V

W

X

Y

MORE TITLES FROM MANNING

Android in Action, Third Edition

by W. Frank Ableson, Robi Sen, Chris King and C. Enrique Ortiz

ISBN: 978-1-617290-50-3
664 pages
$49.99
November 2011

Android in Practice

by Charlie Collins, Michael D. Galpin, Matthias Kaeppler

ISBN: 978-1-935182-92-4
648 pages
$49.99
September 2011

For ordering information go to www.manning.com